ENGINEERING LIBRARY

D0921632

DISCRETE-SIGNAL
ANALYSIS AND DESIGN

BICENTENNIAL
1807
🕮 WILEY
2007
BICENTENNIAL

THE WILEY BICENTENNIAL—KNOWLEDGE FOR GENERATIONS

*E*ach generation has its unique needs and aspirations. When Charles Wiley first opened his small printing shop in lower Manhattan in 1807, it was a generation of boundless potential searching for an identity. And we were there, helping to define a new American literary tradition. Over half a century later, in the midst of the Second Industrial Revolution, it was a generation focused on building the future. Once again, we were there, supplying the critical scientific, technical, and engineering knowledge that helped frame the world. Throughout the 20th Century, and into the new millennium, nations began to reach out beyond their own borders and a new international community was born. Wiley was there, expanding its operations around the world to enable a global exchange of ideas, opinions, and know-how.

For 200 years, Wiley has been an integral part of each generation's journey, enabling the flow of information and understanding necessary to meet their needs and fulfill their aspirations. Today, bold new technologies are changing the way we live and learn. Wiley will be there, providing you the must-have knowledge you need to imagine new worlds, new possibilities, and new opportunities.

Generations come and go, but you can always count on Wiley to provide you the knowledge you need, when and where you need it!

WILLIAM J. PESCE
PRESIDENT AND CHIEF EXECUTIVE OFFICER

PETER BOOTH WILEY
CHAIRMAN OF THE BOARD

DISCRETE-SIGNAL ANALYSIS AND DESIGN

WILLIAM E. SABIN

BICENTENNIAL

1807

WILEY

2007

BICENTENNIAL

WILEY-
INTERSCIENCE

A JOHN WILEY & SONS, INC., PUBLICATION

Copyright © 2008 by John Wiley & Sons, Inc. All rights reserved.

Published by John Wiley & Sons, Inc., Hoboken, New Jersey.
Published simultaneously in Canada.

No part of this publication may be reproduced, stored in a retrieval system, or transmitted in any form or by any means, electronic, mechanical, photocopying, recording, scanning, or otherwise, except as permitted under Section 107 or 108 of the 1976 United States Copyright Act, without either the prior written permission of the Publisher, or authorization through payment of the appropriate per-copy fee to the Copyright Clearance Center, Inc., 222 Rosewood Drive, Danvers, MA 01923, (978) 750-8400, fax (978) 750-4470, or on the web at www.copyright.com. Requests to the Publisher for permission should be addressed to the Permissions Department, John Wiley & Sons, Inc., 111 River Street, Hoboken, NJ 07030, (201) 748-6011, fax (201) 748-6008, or online at http://www.wiley.com/go/permission.

Limit of Liability/Disclaimer of Warranty: While the publisher and author have used their best efforts in preparing this book, they make no representations or warranties with respect to the accuracy or completeness of the contents of this book and specifically disclaim any implied warranties of merchantability or fitness for a particular purpose. No warranty may be created or extended by sales representatives or written sales materials. The advice and strategies contained herein may not be suitable for your situation. You should consult with a professional where appropriate. Neither the publisher nor author shall be liable for any loss of profit or any other commercial damages, including but not limited to special, incidental, consequential, or other damages.

For general information on our other products and services or for technical support, please contact our Customer Care Department within the United States at (800) 762-2974, outside the United States at (317) 572-3993 or fax (317) 572-4002.

Wiley also publishes its books in a variety of electronic formats. Some content that appears in print may not be available in electronic formats. For more information about Wiley products, visit our web site at www.wiley.com.

Wiley Bicentennial Logo: Richard J. Pacifico

Library of Congress Cataloging-in-Publication Data

Sabin, William E.
    Discrete-signal analysis and design / By William E. Sabin.
        p. cm.
    ISBN 978-0-470-18777-7 (cloth/cd)
    1. Signal processing—Digital techniques. 2. Discrete-time systems. 3. System analysis. I. Title.
    TK7868.D5S13 2007
    621.382'2--dc22
                                                                    2007019076

Printed in the United States of America

10 9 8 7 6 5 4 3 2 1

*This book is dedicated to
my wife, Ellen; our sons, Paul and James;
our daughter, Janet; and all of our grandchildren*

CONTENTS

6 Probability and Correlation   95

7 The Power Spectrum   113

PREFACE

The Introduction explains the scope and motivation for the title subject. My association with the Engineering Department of Collins Radio Co., later Rockwell Collins, in Cedar Rapids, Iowa, and my education at the University of Iowa have been helpful background for the topics covered.

The CD accompanying the book includes the Mathcad® V.14 Academic Edition, which is reproduced by permission. This software is fully functional, with no time limitation for its use, but cannot be upgraded. For technical support, more information about purchasing Mathcad, or upgrading from previous editions, see http://www.ptc.com.

Mathcad is a registered trademark of Parametric Technology Corporation (PTC), http://www.ptc.com. PTC owns both the Mathcad software program and its documentation. Both the program and documentation are copyrighted with all rights reserved. No part of the program or its documentation may be produced, transmitted, transcribed, stored in a retrieval system, or translated into any language in any form without the written permission of PTC.

WILLIAM E. SABIN

Introduction

Joseph Fourier 1768-1830

Electronic circuit analysis and design projects often involve time-domain and frequency-domain characteristics that are difficult to work with using the traditional and laborious mathematical pencil-and-paper methods of former eras. This is especially true of certain nonlinear circuits and systems that engineering students and experimenters may not yet be comfortable with.

These difficulties limit the extent to which many kinds of problems can be explored in the depth and as quantitatively as we would like. Specific programs for specific purposes often do not provide a good tie-in with basic principles. In other words, the very important mathematical background and understanding are unclear. Before we can design something we have to look beyond the diagrams, parts lists, and formula handbooks. The reliance on intuitive methods, especially, is becoming increasingly error prone and wasteful.

We can never become too well educated about fundamentals and about our ability to view them from a mathematical perspective. The modern emphasis on math literacy is right on target.

Discrete-Signal Analysis and Design, By William E. Sabin
Copyright © 2008 John Wiley & Sons, Inc.

1

In this book, we will get a better understanding of discrete-time and discrete-frequency signal processing, which is rapidly becoming an important modern way to design and analyze electronics projects of just about every kind. If we understand the basic mathematics of discrete-signal processing sequences, we are off to a good start. We will do all of this at an *introductory* level. The limited goal is to set the stage for the more advanced literature and software, which provide much greater depth. One outstanding example of this is [Oppenheim and Schafer].

What is needed is an easy way to set up a complex problem on a personal computer screen: that is, a straightforward method that provides visual output that is easy to understand and appreciate and illuminates the basic principles involved. Special-purpose personal computer analysis programs exist that are helpful in some of these situations, but they are usually not as useful, flexible, interactive, or easy to modify as the methods that we will explore. In particular, the ability to evaluate easily certain changes in parameter and component values in a problem is a valuable design aid. We do this by interacting with the equations involved. Our approach in this introductory book is almost entirely mathematical, but the level of math is suitable for an undergraduate electrical engineering curriculum and for independent study. Several examples of problems solved in this way are in each of the eight main chapters and Appendix.

By *discrete signals* we mean signals that are in the discrete-time $x(n)$ and discrete-frequency $X(k)$ domains. Amplitude values are continuous. This differs from digital signal processing (DSP), which is also discrete (quantized) in amplitude. With personal computers as tools, the persons who use them for various activities, especially electronic engineering activities, are especially comfortable with this approach, which has become highly developed. The math is especially practical. Discrete signals are a valuable middle ground between classical-continuous and DSP.

In an electronics lab, data points are almost always obtained (very often automatically) at discrete values and discrete intervals of time and frequency. The discrete methods of this book are therefore very practical ways to analyze and process discrete data.

The Discrete Fourier Transform (DFT) and its inverse (IDFT) are the simple tools that convert the information back and forth between the discrete-time and discrete-frequency domains. The Fast Fourier

Transform and its in inverse (IFFT) are the high-speed tools that can expedite these operations. Convolution, correlation, smoothing, windowing, spectral leakage, aliasing, power spectrum, Hilbert transform, and other kinds of sequence manipulations and processing will be studied. We also look for legitimate simplifications and assumptions that make the process easier, and we practice the "art" of approximation. The simplicity of this discrete approach is also the source of its elegance.

Keep in mind that this book deals only with non-real-time analysis and is not involved with high-speed real-time processing. This helps to define our limited tutorial objective.

Be aware also that this book cannot get into the multitude of advanced analytical or experimental methods of lumped or distributed circuits and systems that tell us how a particular signal sequence is obtained: for example, by solutions of differential equations or system analysis. One brief exception to this is in the Appendix. The vast array of literature does all of this much better in specific situations. We assume that the waveforms have been measured or calculated in discrete sequence form as a function of time or frequency. Sampling methods and computer add-on modules are available that do this quite well in the lab at modest cost.

Another important point is that a discrete sequence does not always have some particular defining equation that we can recognize. It can very easily be experimental data taken from lab measurements, from published graphs or tables, from a set of interconnected segments, or just simply something that is imagined or "what if we try this?" It can be random or pseudorandom data that we want to analyze or process. The data can be in time domain or frequency domain, and we can easily move the data and the results back and forth between those domains. For example, a noise-contaminated spectrum can be filtered in various ways, and the results can be seen in the time domain. The noisy time domain-to-frequency domain conversion results can also be seen easily.

A basic assumption for this book is that a discrete signal sequence from 0 to N-1 in the time or frequency domain is just one segment of an infinitely repeating steady-state sequence. Each sequence range contains all of the significant time and frequency content that we need in order to get a "reasonable" approximation that can stand alone. We design and process the segment and its length N so that this condition is sufficiently satisfied. A further assumption is that a sequence contains a positive time or frequency part and an equal-length negative time or frequency part.

MATHCAD®

I have thought a great deal about the best way to perform the mathematical operations that are to be discussed. In these modern times, an easy-to-use and highly regarded math program such as my personal preference, Mathcad (Parametric Technology Corporation, www.ptc.com), that can perform complex and nonlinear math operations of just about any kind, has become very popular. The equations and functions are typed directly onto the computer screen "writing tablet" or "blackboard" (a.k.a "whiteboard") in math-book format [Barry Simon]. A relatively easy learning process gets us started; however, familiarity with Mathcad's rules and regulations does need some time, just like any new software that we encounter. The simplicity and user friendliness are easy to appreciate. Mathcad is very sophisticated, but in this book we will only need to scratch the surface.

A special one-purpose program written in a tedious programming language that works only with a single project does not make nearly as much sense as a more versatile software that quickly and easily serves a wide variety of projects and purposes for many years. Mathcad does that very well, and the results can be archived "forever." A dedicated special program just doesn't have the same versatility to handle easily the special situations which, for most of us, happen very often. Mathcad is excellent for persons who do not want to become deeply involved with structured languages.

A significant advantage of Mathcad is the ease and speed with which the equations, parameters, data, and graphs can be modified in an experimental mode. Also, having all of this basic information in front of our eyes is a powerful connection to the mathematics. With structured languages we are always creating programming language linkages, with all of their syntax baggage, between the problem and the result. We are always parsing the lines of code to figure out what is going on. Working directly with the math, in math format, greatly reduces all of that. In short, Mathcad is a relatively pleasant interactive calculation program for applied math projects.

However, it is important to point out also that this book is not an instruction manual for Mathcad. The Mathcad *User Guide* and the very complete and illustrated Help (F1) section do that much better than I can. We will use Mathcad at its introductory level to help us understand the basic principles of discrete-signal processing, which is our only goal. Learning experience will lead to greater proficiency. One of Mathcad's useful tools is the "Ctrl Z", which can "undo" one or many incorrect keystrokes.

Classroom versions of Mathcad are available but ordinarily require a Student Authorization. The only limitation to the special Student Version is that it cannot be upgraded at low cost to later standard versions of Mathcad.

The latest standard version, purchased new, although a significant initial expense, is an excellent long-term resource and a career investment for ·the technically oriented individual with mathematical interests, and the occasional future version upgrades are inexpensive. The up-front cost of the Mathcad standard version compares quite favorably with competitive systems, and is comparable in terms of features and functionality. The standard version of Mathcad is preferable, in my opinion.

There is embedded in Mathcad a "Programming Language" capability that is very useful for many applications. The Help (F1) guide has some very useful instructions for "Programming" that help us to get started. These programs perform branching, logical operations, and conditional loops, with embedded complex-valued math functions and Mathcad calculations of just about any type. This capability greatly enhances Mathcad's usefulness. This book will show very simple examples in several chapters.

A complete, full-featured copy of Mathcad, with unlimited time usage, accompanies this book. It should ethically not be distributed beyond the initial owner.

It is also important to point out that another software approach, such as MATLAB®, is an excellent alternate when available. In fact, Mathcad interacts with MATLAB in ways that the Mathcad *User Guide* illustrates. My experience has been that with a little extra effort, many MATLAB functions revert to Mathcad methods, especially if the powerful *symbolic math* capabilities of Mathcad are used. MATLAB users will have no trouble translating everything in this book directly to their system. Keep printouts and notes for future reference. Mathcad also has an excellent relationship with an EXCEL program that has been configured for complex algebra. EXCEL is an excellent partner to Mathcad for many purposes.

An excellent, high-quality linear and nonlinear analog and digital circuit simulator such as Multisim (Electronics Work Bench, a division of world-famous National Instruments Co., www.ni.com), which uses accurate models for a wide range of electronic components, linear and nonlinear, is another long-term investment for the serious electronics engineer and experimenter. And similar to Mathcad, your circuit diagram, with component values and many kinds of virtual test instruments, appears on the screen. A sophisticated embedded graphing capability is included. Less expensive (or even free) but fairly elementary alternatives are available from many other sources. For example, the beginner may want to start with the various forms of SPICE. However, Multisim, although the up-front cost is significant, is a valuable long-term investment that should be considered. Multisim offers various learning editions at reduced cost. I recommend this software, especially the complete versions, very highly as a long-term tool for linear and nonlinear analysis and simulation. An added RF Design package is available for more sophisticated RF modelling.

Mathcad is also interactive with LabVIEW, another product of National Instruments Co., which is widely used for laboratory data gathering and analysis. See http://www.ni.com/analysis/mathcad.htm for more information on this interesting topic.

Another approach that is much less expensive, but also much less powerful, involves structured programming languages such as BASIC, Fortran, C^{++}, Pascal, EXCEL, and others with which many readers have previous experience. However, my suggestion is to get involved early with a more sophisticated and long-enduring approach, especially with an excellent program such as Mathcad.

For the website-friendly personal computer, the online search engines put us in touch very quickly with a vast world of specific technical reference and cross-referenced material that would often be laborious to find using traditional library retrieval methods.

MathType, an Equation Editor for the word processor (http://www.dessci.com/en/), is another valuable tool that is ideal for document and report preparation. This book was written using that program.

And of course these programs are all available for many other uses for many years to come. The time devoted to learning these programs, even at the introductory level, is well spent. These materials are not free, but in my opinion, a personal at-home modest long-term investment in productivity software should be a part of every electronics engineer's and experimenter's career (just like his education), as a supplement to that which is at a school or company location (which, as we know, can change occasionally).

Keep in mind that although the computer is a valuable tool, it does not relieve the operator of the responsibility for understanding the core technology and math that are being utilized. Nevertheless, some pleasant and unexpected insights will occur very often.

Remember also that the introductory treatment in this book is not meant to compete with the more scholarly literature that provides much more advanced coverage, but hopefully, it will be a good and quite useful initial contact with the more advanced topics.

REFERENCES

Oppenheim, A. V., and R. W., Schafer, 1999, *Discrete-Time Signal Processing*, 2nd ed., Prentice Hall, Upper Saddle River, NJ.

Simon, B., Various Mathcad reviews, Department of Mathematics, California Institute of Technology.

1

First Principles

This first chapter presents an overview of some basic ideas. Later chapters will expand on these ideas and clarify the subtleties that are frequently encountered. Practical examples will be emphasized. The data to be processed is presented in a sampled-time or sampled-frequency format, using a number of samples that is usually not more than $2^{11} = 2048$. The following "shopping list" of operations is summarized as follows:

1. The user inputs, from a tabulated or calculated sequence, a set of numerical values, or possibly two sets, each with $N = 2^M (M = 3, 4, 5, \ldots ,11)$ values. The sets can be real or complex in the "time" or "frequency" domains, which are related by the Discrete Fourier Transform (DFT) and its companion, the Inverse Discrete Fourier Transform (IDFT). This book will emphasize time and frequency domains as used in electronic engineering, especially communications. The reader will become more comfortable and proficient in both domains and learn to think simultaneously in both.

2. The sequences selected are assumed to span one period of an eternal steady-state repetitive sequence and to be highly separated from

Discrete-Signal Analysis and Design, By William E. Sabin
Copyright © 2008 John Wiley & Sons, Inc.

adjacent sequences. The DFT (discrete Fourier transform), and DFS (discrete Fourier series) are interchangeable in these situations.

3. The following topics are emphasized:

  a. Forward transformation and inverse transformation to convert between "frequency" and "time".

  b. Spectral leakage and aliasing.

  c. Smoothing and windowing operations in time and frequency.

  d. Time and frequency scaling operations.

  e. Power spectrum and cross-spectrum.

  f. Multiplication and convolution using the DFT and IDFT.

  g. Relationship between convolution and multiplication.

  h. Autocorrelation and cross-correlation.

  i. Relations between correlation and power spectrum using the Wiener-Khintchine theorem.

  j. Filtering or other signal-processing operations in the time domain or frequency domain.

  k. Hilbert transform and its applications in communications.

  l. Gaussian (normal) random noise.

  m. The discrete differential (difference) equation.

The sequences to be analyzed can be created by internal algorithms or imported from data files that are generated by the user. A library of such files, and also their computed results, can be named and stored in a special hard disk folder.

The DFT and IDFT, and especially the FFT and IFFT, are not only very fast but also very easy to learn and use. Discrete Signal Processing using the computer, especially the personal computer, is advancing steadily into the mainstream of modern electrical engineering, and that is the main focus of this book.

SEQUENCE STRUCTURE IN THE TIME AND FREQUENCY DOMAINS

A time-domain sequence $x(n)$ of infinite duration $-\infty \leq n \leq +\infty$ that repeats at multiples of N is shown in Fig. 1-1a, where each $x(n)$ is uniquely

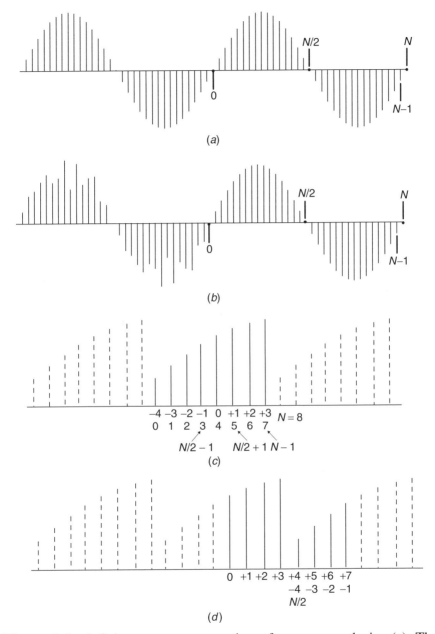

(a)

(b)

(c)

(d)

Figure 1-1 Infinite sequence operations for wave analysis. (a) The segment of infinite periodic sequence from 0 to $N - 1$. The next sequence starts at N. (b) The Segment of infinite sequence from 0 to $N - 1$ is not periodic with respect to the rest of the infinite sequence. (c) The two-sided sequence starts at $- 4$ or 0. (d) The sequence starts at 0.

identified in both time and amplitude. If the sequence is nonrepeating (random), or if it is infinite in length, or if it is periodic but the sequence is not chosen to be exactly one period, then this segment is not one period of a truly periodic process, as shown in Fig. 1-1b. However, the wave analysis math assumes that the part of the wave that is selected is actually periodic within an infinite sequence, similar to Fig. 1-1a. The selected sequence can then perhaps be referred to as "pseudo-periodic", and the analysis results are correct for that sequence. For example, the entire sequence of Fig. 1-1b, or any segment of it, can be analyzed exactly as though the selected segment is one period of an infinite periodic wave. The results of the analysis are usually different for each different segment that is chosen. If the 0 to $N - 1$ sequence in Fig. 1-1b is chosen, the analysis results are identical to the results for 0 to $N - 1$ in Fig. 1-1a.

When selecting a segment of the data, for instance experimentally acquired values, it is important to be sure that the selected data contains the amount of information that is needed to get a sufficiently accurate analysis. If amplitude values change significantly between samples, we must use samples that are more closely spaced. There is more about this later in this chapter.

It is important to point out a fact about the time sequences $x(n)$ in Fig. 1-1. Although the samples are shown as thin lines that have very little area, each line does represent a definite amount of energy. The sum of these energies, within a unit time interval, and if there are enough of them so that the waveform is adequately represented (the Nyquist and Shannon requirements) [Stanley, 1984, p. 49], contains very nearly the same energy per unit time interval; in other words very nearly the same average power (theoretically, exactly the same), as the continuous line that is drawn through the tips of the samples [Carlson, 1986, pp. 351 and 624]. Another way to look at it is to consider a single sample at time (n) and the distance from that sample to the next sample, at time $(n + 1)$. The area of that rectangle (or trapezoid) represents a certain value of energy. The value of this energy is proportional to the length (amplitude) of the sample. We can also think of each line as a Dirac "impulse" that has zero width but a definite area and an amplitude $x(n)$ that is a measure of its energy. Its Laplace transform is equal to 1.0 times $x(n)$.

If the signal has some randomness (nearly all real-world signals do), the conclusion of adequate sampling has to be qualified. We will see in later chapters, especially Chapter 6, that one record length (N) of such a signal may not be adequate, and we must do an averaging operation, or other more elaborate operations, on many such records.

Discrete sequences can also represent samples in the frequency domain, and the same rules apply. The power in the adequate set of individual frequencies over some specified bandwidth is almost (or exactly) the same as the power in the continuous spectrum within the same bandwidth, again assuming adequate samples.

In some cases it will be more desirable, from a visual standpoint, to work with the continuous curves, with this background information in mind. Figure 1-6 is an example, and the discrete methods just mentioned are assumed to be still valid.

TWO-SIDED TIME AND FREQUENCY

An important aspect of a periodic time sequence concerns the relative time of occurrence. In Fig. 1-1a and b, the "present" item is located at $n = 0$. This is the reference point for the sequence. Items to the left are "previous" and items to the right are "future". Figure 1-1c shows an 8-point sequence that occurs between -4 and $+3$. The "present" symbol is at $n = 0$, previous symbols are from -4 to -1, and future symbols are from $+1$ to $+3$. In Fig. 1-1d the same sequence is shown labeled from 0 to $+7$. But the $+4$ to $+7$ values are observed to have the same amplitudes as the -4 to -1 values in Fig. 1-1c. Therefore, the $+4$ to $+7$ values of Fig. 1-1d should be thought of as "previous" and they may be relabeled as shown in Fig. 1-1d. We will use this convention consistently throughout the book. Note that one location, $N/2$, is labeled both as $+4$ and -4. This location is special and will be important in later work. In computerized waveform analysis and design, it is a good practice to use $n = 0$ as a starting point for the sequence(s) to be processed, as in Fig. 1-1d, because a possible source of confusion is eliminated.

A similar but slightly different idea occurs in the frequency-domain sequence, which is usually a two-sided spectrum consisting of positive- and negative-frequency harmonics, to be discussed in detail later. For example, if Fig. 1-1c and d are frequency values $X(k)$, then -4 to -1 in Fig. 1-1c and $+4$ to $+7$ in Fig. 1-1d are negative frequencies. The value at

$k = 0$ is the dc component, $k = \pm 1$ is the $\pm$*fundamental* frequency, and other $\pm k$ values are $\pm$*harmonics* of the $k = \pm 1$ value. The frequency $k = \pm N/2$ is special, as discussed later. Because of the assumed steady-state periodicity of the sequences, the Discrete Fourier Transform, often *correctly* referred to in this book as the Discrete Fourier Series, and its inverse transform are used to travel very easily between the time and frequency domains.

An important thing to keep in mind is that in all cases, in this chapter or any other where we perform a summation (Σ) from 0 to $N - 1$, we assume that all of the *significant* signal and noise energy that we are concerned with lies within those boundaries. We are thus relieved of the integrations from $-\infty$ to $+\infty$ that we find in many textbooks, and life becomes simpler in the discrete 0 to $N - 1$ world. It also validates our assumptions about the steady-state repetition of sequences. In Chapters 3 and 4 we look at aliasing, spectral leakage, smoothing, and windowing, and these help to assure our reliance on 0 to $N - 1$. We can also increase N by $2^M (M = 2,$ 3, 4, ...) as needed to encompass more time or more spectrum.

DISCRETE FOURIER TRANSFORM (SERIES)

A typical example of discrete-time $x(n)$ values is shown in Fig. 1-2a. It consists of 64 equally spaced real-valued samples $0 \leq n \leq 63$ of a sine wave, peak amplitude $A = 1.0$ V, to which a dc bias of Vdc $= + 1.0$ V has been added. Point $n = N = 64$ is the beginning of the next sine wave plus dc bias. The sequence $x(n)$, including the dc component, is

$$x(n) = A \sin\left(2\pi \frac{n}{N} K_x\right) + \text{Vdc} \qquad \text{volts} \qquad (1\text{-}1)$$

where K_x is the number of cycles per sequence length: in this example, 1.0. To find the frequency spectrum $X(k)$ for this $x(n)$ sequence (Fig. 1-2b), we use the DFT of Eq. (1-2) [Oppenheim et al., 1983, p. 321]:

$$X(k) = \frac{1}{N} \sum_{n=0}^{N-1} x(n)\, e^{-j2\pi \frac{n}{N} \cdot k} \qquad \text{volts}, \qquad k = 0 \text{ to } N - 1 \quad (1\text{-}2)$$

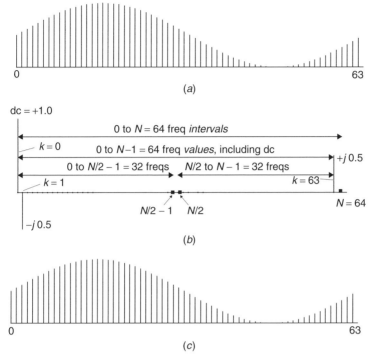

Figure 1-2 Sequence (a) is converted to a spectrum (b) and reconverted to a sequence (c). (a) 64-point sequence, sine wave plus dc bias.
(b) Two-sided spectrum of w to count freq part (a) showing ho values
and frequency intervals. (c) The spectrum of part (b) is reconverted to the
time sequence of part (a).

In this equation, for each discrete value of (k) from 0 to $N - 1$, the function $x(n)$ is multiplied by the complex exponential, whose magnitude $=$
1.0. Also, at each (n) a constant negative (clockwise) phase lag increment $(-2\pi nk/N)$ radians is added to the exponential. Figure 1-2b shows
that the spectrum has just two lines of amplitude $\pm j\,0.5$ at $k = 1$ and 63,
which is correct for a sine wave of frequency 1.0, plus the dc at $k = 0$.

These two lines combine coherently to produce a real sine wave of
amplitude $A = 1.0$. The peak power in a 1.0 ohm resistor is not the sum of
the peak powers of the two components, which is $(0.5^2 + 0.5^2) = 0.5\,\text{W}$;
instead, the peak power is the square of the sum of the two components,
which is $(0.5 + 0.5)^2 = 1.0\,\text{W}$. If the spectrum component $X(k)$ has a real

part and an imaginary part, the real parts add coherently and the imaginary parts add coherently, and the power is complex (real watts and imaginary vars). There is much more about this later.

If $K_x = 1.2$ in Eq. (1-1), then 1.2 cycles would be visible, the spectrum would contain many frequencies, and the final phase would change to $(0.2 \cdot 2\pi)$ radians. The value of the phase angle in degrees for each complex $X(k)$ is

$$\phi(k) = \arctan\left(\frac{\text{Im}\left(X(k)\right)}{\text{Re}\left(X(k)\right)}\right) \cdot \frac{180}{\pi} \qquad \text{degrees} \qquad (1\text{-}3)$$

For an example of this type of sequence, look ahead to Fig. 1-6. A later section of this chapter gives more details on complex frequency-domain sequences.

At this point, notice that the complex term $\exp(j\omega t)$ is calculated by Mathcad using its powerful and efficient algorithms, eliminating the need for an elaborate complex Taylor series expansion by the user at each value of (n) or (ω). This is good common sense and does not derail us from our discrete time/frequency objectives.

At each (k) stop, the sum is performed at 0 to $N - 1$ values of time (n), for a total of N values. It may be possible to evaluate accurately enough the sum at each (k) value with a smaller number of time steps, say $N/2$ or $N/4$. For simplicity and best accuracy, N will be used for both (k) and (n). Using Mathcad to find the spectrum without assigning discrete (k) values from 0 to $N - 1$, a very large number of frequency values are evaluated and a continuous graph plot is created. We will do this from time to time, and the summation (Σ) becomes more like an integral $\left(\int\right)$, but this is not always a good idea, for reasons to be seen later.

Note also that in Eq. (1-2) the factor $1/N$ ahead of the sum and the minus sign in the exponent are used but are not used in Eq. (1-8) (look ahead). This notation is common in engineering applications as described by [Ronald Bracewell, 1986] and is also an option in Mathcad (functions FFT and IFFT). See also [Oppenheim and Willsky et al., 1983, p. 321]. This agrees with the practical engineering emphasis of this book. It also agrees with our assumption that each record, 0 to $N - 1$, is one replication of an infinite steady-state signal. These two equations, used together and consistently, produce correct results.

Each (k) is a harmonic number for the frequency sequence $X(k)$. To repeat a few previous statements for emphasis, $k = 1$ is the fundamental frequency, $k = 2$ is second harmonic, etc. A two-sided (positive and negative) phasor spectrum is produced by this equation (we will learn to appreciate the two-sided spectrum concept). N, an integral power of 2, is chosen large enough to provide adequate resolution of the spectrum (sufficient harmonics of $k = 1$). The dc component is at $k = 0$ [where the exp(0) term $= 1.0$] and

$$X(0) = \frac{1}{N} \sum_{n=0}^{N-1} x(n) = \langle x(n) \rangle \qquad \text{volts} \qquad (1\text{-}4)$$

which is the *time average* over the entire sequence, 1.0, in Fig. 1-2.

Equation (1-2) can be used directly to get the spectrum, but as a matter of considerable interest later it can be separated into two regions having an equal number of data points, from 0 to $N/2 - 1$ and from $N/2$ to $N - 1$ as shown in Eq. (1-5). If $N = 8$, then k (positive frequencies) $= 1, 2, 3$ and k (negative frequencies) $= 7, 6, 5$. Point N is the beginning of the next periodic continuation. Dc is at $k = 0$, and $N/2$ is not used, for reasons to be explained later in this chapter.

Consider the following manipulations of Eq. (1-2):

$$X(k) = \frac{1}{N} \left[\sum_{n=0}^{N/2-1} x(n) e^{-jk2\pi\left(\frac{n}{N}\right)} + \sum_{n=N/2}^{N-1} x(n) e^{-jk2\pi\left(\frac{n}{N}\right)} \right] \qquad (1\text{-}5)$$

The last exponential can be modified as follows without changing its value:

$$e^{-jk2\pi\frac{n}{N}} = \underbrace{e^{j(2\pi n)}}_{360°} e^{-jk2\pi\frac{n}{N}} = e^{j2\pi n\left(1-\frac{k}{N}\right)} = e^{j2\pi(N-k)\frac{n}{N}} \qquad (1\text{-}6)$$

and Eq. (1-2) becomes

$$X(k) = \frac{1}{N} \left[\sum_{n=0}^{N/2-1} x(n) e^{-jk2\pi\frac{n}{N}} + \sum_{n=N/2}^{N-1} x(n) e^{j2\pi(N-k)\frac{n}{N}} \right] \qquad (1\text{-}7)$$

The second exponential is the phase conjugate $(e^{-j\theta} \rightarrow e^{+j\theta})$ of the first and is positioned as shown in Fig. 1-2b for $k = N/2$ to $N - 1$. At $k = 0$ we see the dc. The two imaginary components $-j0.5$ and $+j0.5$, are at $k = 1$ and $k = 63$ (same as $k = -1$), typical for a sine wave of length 64. We use this method quite often to convert two-sided sequences into one-sided (positive-time or positive-frequency) sequences (see Chapter 2 for more details).

INVERSE DISCRETE FOURIER TRANSFORM

The inverse transformation (IDFT) in Eq. (1-8) [Oppenheim et al., 1983, p. 321] takes the two-sided spectrum $X(k)$ in Fig. 1-2b and exactly recreates the original two-sided time sequence $x(n)$ shown in Fig. 1-2c:

$$x(n) = \sum_{k=0}^{N-1} X(k)e^{jk2\pi\left(\frac{n}{N}\right)} \tag{1-8}$$

At each value of (n) the spectrum values $X(k)$ are summed from $k = 0$ to $k = N - 1$. In Eq. (1-8) the phase increments are in the counter-clockwise (positive) direction. This reverses the negative phase increments that were introduced into the DFT [Eq. (1-2)]. This step helps to return each *complex* $X(k)$ in the frequency domain to a *real* $x(n)$ in the time domain. See further discussion later in the chapter.

It is interesting to focus our attention on Eqs. (1-2) and (1-8) and to observe that in both cases we are simultaneously in the time and frequency domains. We must have data from both domains to travel back and forth. This confirms that we are learning to be comfortable in both domains at once, which is exactly what we need to do.

So far, Eqs. (1-2) and (1-8) have been used directly, without any need for a faster method, the FFT (the Fast Fourier Transform), described later. Modern personal computers are usually fast enough for simple problems using just these two equations. Also, Eqs. (1-2) and (1-8) are quite accurate and very easy to use in computerized analysis (however, Mathcad also has very excellent tools for numerical and symbolic integration that we will use frequently). We do not have to worry about those two discrete equations in our applications because they have been thoroughly tested. It is a good idea to use Eqs. (1-2) and (1-8) together as a pair. To narrow the time or frequency resolution, multiply the value of N by $2^M (m = 1, 2, 3, \ldots)$, as shown in the next section.

FREQUENCY AND TIME SCALING

Suppose a signal spectrum extends from 0 Hz to 30 MHz (Fig. 1-3) and we want to display it as a 32-point ($=2^5$) two-sided spectrum. The positive side of the spectrum has 15 $X(k)$ values from 1 to $N/2 - 1$ (not counting 0 and $N/2$), and the negative side of the spectrum also has 15 $X(k)$ values from $N/2 + 1$ to $N - 1$ (not counting $N/2$ and N). The frequency range 0 to 30 MHz consists of a fundamental frequency k_1 and $2^4 - 1 = 15$ harmonics of k_1. The fundamental frequency $k1$ is determined by

$$k_1 \cdot 15 = 3 \cdot 10^7 \quad \therefore \quad k_1 = \frac{3 \cdot 10^7}{15} = 2 \, \text{MHz} \tag{1-9}$$

and this is the best resolution of frequency that can be achieved with 15 points (positive or negative frequencies) of a 30-MHz signal using a 32-point two-sided spectrum. If we use 2048 data points, we can get 29.31551-kHz resolution using Eq. (1-9).

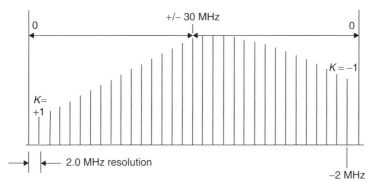

Figure 1-3 A 30-MHz two-sided spectrum with 32 frequency samples, including 0.

An excellent way to improve this example is to frequency-convert the signal band to a much lower frequency, for example 3 MHz, using a very stable local oscillator, which would give us a 2931.55-Hz resolution for this example. Increasing the samples to 2^{14} at 3 MHz provides a resolution of 366.26 Hz, and so forth for higher sample numbers. This is basically what spectrum analyzers do.

The good news for this problem is that a hardware frequency translator may not be necessary. If the signal is narrowband, such as speech or low-speed data or some other bandlimited process, the original 30-MHz problem might be restated at 3 MHz, or maybe even at 0.3 MHz, with the same signal bandwidth and with no loss of correct results, but with greatly improved resolution. With programs for personal computer analysis, very large numbers of samples are not desirable; therefore, we do not try to push the limits too much. The waveform analysis routines usually tell us what we want to know, using more reasonable numbers of samples. Designing the frequency and time scales is very helpful.

Consider a time scaling example, a sequence (record length) that is 10 μsec long from start of one sequence to the start of the next sequence, as shown in Fig. 1-4. For $N = 4$ there are 4 time *values* (0, 1, 2, 3) and 4 time *intervals* (1, 2, 3, 4) to the beginning of the next sequence, which is $10^{-5}/4 = 2.5$ μsec per interval. In the first half there are 2 intervals for a total of 5.0 μsec. For the second half there are also 2 intervals, for a total of 5.0 μsec. Each interval is a "band" of possibly smaller time increments. The total time is 10.0 μsec.

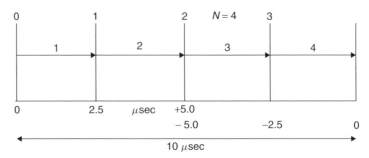

Figure 1-4   A 10-μsec time sequence with positive and negative time values.

For $N = 2^M$ points there are N *values*, including 0, and N *intervals* to the beginning of the next sequence. For a two-sided time sequence the special midpoint term $N/2$ can be labeled as $+5.0$ μsec and also -5.0 μsec, as shown in Fig. 1-4. It is important to do this time scaling correctly.

Figure 1-2b shows an identical way to label frequency values and frequency intervals. Each value is a specific frequency and each interval is a frequency "band". This approach helps us to keep the spectrum more clearly in mind. If amplitude values change too much within an interval, we will use a higher value of N to improve frequency resolution, as discussed previously. The same idea applies in the time domain. The term *picket fence effect* describes the situation where the selected number of integer values of frequency or time does not give enough detail. It's like watching a ball game through a picket fence.

NUMBER OF SAMPLES

The sampling theorem [Carlson, 1986, p. 351] says that a single sine wave needs more than two, preferably at least three, samples per cycle. A frequency of 10,000 Hz requires $1/(10,000 \cdot 3) = 3.33 \cdot 10^{-5}$ seconds for each sample. A signal at 100 Hz needs $1/(100 \cdot 3) = 3.33 \cdot 10^{-3}$ seconds for each sample. If both components are present in the same composite signal, the minimum required total number of samples is $(3.33 \cdot 10^{-3})/(3.33 \cdot 10^{-5}) = 10^2 = 100$. In other words, 100 cycles of the 10,000-Hz component occupy the same time as 1 cycle of the 100-Hz component. Because the time sequence is two-sided, positive time and negative time, 200 samples would be a better choice. The nearest preferred value of N is $2^8 = 256$, and the sequence is from $0 \le n \le N - 1$. The plot of the DFT phasor spectrum $X(k)$ is also two-sided with 256 positions. $N = 256$ is a good choice for both time and frequency for this example.

If a particular waveform has a well-defined time limit but insufficient nonzero data values, we can improve the time resolution and therefore the frequency resolution by adding *augmenting zeros* to the time-domain data. Zeros can be added before and after the limited-duration time signal. The total number of points should be $2^M (M = 2, 3, 4, \ldots)$, as mentioned before. Using Eq. (1-8) and recalling that a time record N produces $N/2$

positive-frequency phasors and $N/2$ negative-frequency phasors, the frequency resolution improves by the factor (total points)/(initial points). The spectrum can sometimes be distorted by this procedure, and *windowing* methods (see Chapter 4) can often reduce the distortion.

COMPLEX FREQUENCY DOMAIN SEQUENCES

We discuss further the complex frequency domain $X(k)$ and the phasor concept. This material is very important throughout this book.

The complex plane in Fig. 1-5 shows the locus of imaginary values on the vertical axis and the locus of real values on the horizontal axis. The directed line segment Ae^{je}, also known as a *phasor*, especially in electronics, has a horizontal (real) component $A\cos\theta$ and a vertical (imaginary) component $jA\sin\theta$. The phasor rotates counter-clockwise at a positive angular rate (radians per second) $= 2\pi f$. At the *frozen* instant of time in the diagram the phase *lead* of phasor 1 relative to phasor 2 becomes $\theta = \omega\Delta t = 2\pi f\Delta t$. That is, phasor 1 will reach its maximum amplitude (in the vertical direction) *sooner* than phasor 2 therefore, *phasor 1 leads phasor 2* in phase and also in time. A time-domain sine-wave diagram of phasor 1 and 2 verifies this logic. We will see this again in Chapter 5.

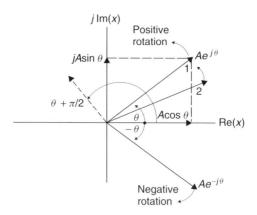

Figure 1-5   Complex plane and phasor example.

The letter j has dual meanings: (1) it is a mathematical *operator*,

$$e^{j\pi/2} = \cos\left(\frac{\pi}{2}\right) + j\sin\left(\frac{\pi}{2}\right) = 0 + j1 = j \qquad (1\text{-}10)$$

that performs a $90°$ (*quadrature*) counter-clockwise *leading* phase shift on any phasor in the complex plane, for example from $45°$ to $135°$, and (2) it is used as a *label* to tell us that the quantity following it is on the imaginary axis: for example, $R + jX$, where R and X are both real numbers. The *conjugate* of the phase-leading phasor at angle (θ) is the phase-lagging clockwise-rotating phasor at angle ($-\theta$). The *quadrature* angle is $\theta \pm 90°$.

TIME *x(n)* VERSUS FREQUENCY *X(k)*

It is very important to keep in mind the concepts of two-sided time and two-sided frequency and also the idea of complex-valued *sequences* $x(n)$ in the time domain and complex-valued *samples* $X(k)$ in the frequency domain, as we now explain.

There is a distinction between a sample in time and a sample in frequency. An individual time sample *x(n)*, where we define x to be a real number, has two attributes, an amplitude value x and a time value (n). There is no "phase" or "frequency" associated with this $x(n)$, *if viewed by itself*. A special clarification to this idea follows in the next paragraph. Think of the $x(n)$ *sequence* as an oscilloscope screen display. This sequence of time samples may have some combination of frequencies and phases that are defined by the variations in the amplitude and phase of the sequence. The DFT in Eq. (1-2) is explicitly designed to give us that information by examining the time sequence. For example, a phase change of the entire sequence slides the entire sequence left or right. A sine wave sequence in phase with a $0°$ reference phase is called an (*I*) wave and a sine wave sequence that is at $90°$ with respect to the (*I*) wave sequence is called a (*Q* or *jQ*) quadrature wave. Also, an individual time sample *x(n)* can have a "phase identifier" by virtue of its position in the time sequence. So we may speak in this manner of the phase and frequency of an $x(n)$ time sequence, but we must avoid confusion on this issue. In this book, each $x(n)$ in the time domain is assumed to be a "real" signal, but the "wave" may be complex in the sense that we have described.

A special circumstance can clarify the conclusions in the previous paragraph. Suppose that instead of $x(n)$ we look at $x(n)\exp(j\theta)$, where θ is a *constant* angle as suggested in Fig. 1-5. Then (see also p. 46)

$$x(n)\exp(j\theta) = x(n)\cos(\theta) + jx(n)\sin\theta = I(n) + jQ(n) \qquad (1\text{-}11)$$

and we now have two *sequences* that are in phase quadrature, and each sequence has real values of $x(n)$. Finally, suppose that the constant θ is replaced by the time-varying $\theta(n)$ from $n = 0$ to $N - 1$. Equation (1-11) becomes $x(n)\exp[j\theta(n)]$, which is a *phase modulation* of $x(n)$. If we plug this into the DFT in Eq. (1-2) we get the spectrum

$$
\begin{aligned}
X(k) &= \frac{1}{N}\sum_{n=0}^{N-1}\left[x(n)\exp[j\theta(n)]\right]\exp\left(-j2\pi\frac{n}{N}k\right)\\
&= \frac{1}{N}\sum_{n=0}^{N-1}x(n)\exp\left\{-j\left[2\pi\frac{n}{N}k - \theta(n)\right]\right\}
\end{aligned}
\qquad (1\text{-}12)
$$

where k can be any value from 0 to $N - 1$ and the time variations in $\theta(n)$ become part of the spectrum of a phase-modulated signal, along with the part of the spectrum that is due to the *peak* amplitude variations (if any) of $x(n)$. Equation (1-12) can be used in some interesting experiments. Note the ease with which Eq. (1-12) can be calculated in the discrete-time/frequency domains. In this book, in the interest of simplicity, we will assume that the $x(n)$ values are real, as stated at the outset, and we will complete the discussion.

A frequency *sample* $X(k)$, which we often call a *phasor*, is also a voltage or current value X, but it also has *phase* $\theta(k)$ relative to some reference θ_R, and *frequency* k as shown on an $X(k)$ graph such as Fig. 1-2b, $k = +1$ and $k = +63$ (same as -1). The phase angle $\theta(k)$ of each phasor can

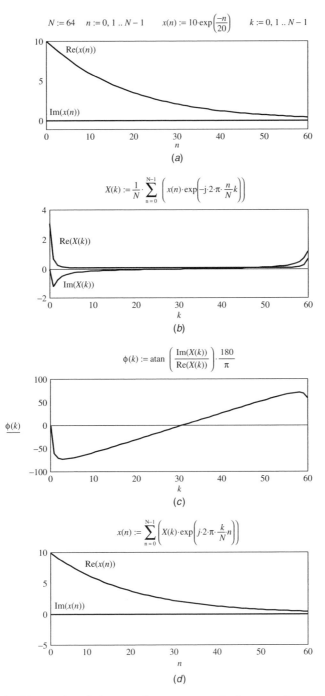

Figure 1-6   Example of time to frequency and phase and return to time.

be shown, if we like, on a separate phase-angle graph (Fig. 1-6). Finally, to reconstruct the time plot in Fig. 1-2c, the two rotating $X(k)$ phasors in Fig. 1-2b re-create the sinusoidal time sequence $x(n)$, using the IDFT of Eq. (1-8). Figure 1-6 should be studied as an example of converting $\exp(-n/20)$ from time to frequency and phase and back to time. Note that parts (a) and (d) show only the positive-time part of the $x(n)$ waveform. The negative-time part is a mirror image and is occasionally not shown, but it is never ignored.

There is one other thing about sequences. Because in this book they are steady-state signals in which all transients have disappeared, it does not matter where they came from. They can be solutions to differential equations, or signal generator output at the end of a long nonlinear transmission line, etc., etc. The DFT and IDFT do not identify the source of the sequences, only tell the relationship between the steady-state time domain and the steady-state frequency domain. We should avoid trying to make anything more than that out of them. Other methods do a much better job of tracing the *origins* of sequences in time and frequency. The Appendix shows a simple example of this interesting and very important activity.

REFERENCES

Bracewell, R., 1986, *The Fourier Transform and its Applications*, McGraw-Hill, New York.

Carlson, A. B., 1986, *Communication Systems*, 3rd ed., McGraw-Hill, New York.

Oppenheim, A.V., A. Willsky, and I. Young, 1983, *Signals and Systems*, Prentice–Hall, Englewood Cliffs, NJ.

Stanley, W. D., et al., 1984, *Digital Signal Processing*, 2nd ed., Reston Publishing, Reston, VA.

2

Sine, Cosine, and θ

In the spectrum $X(k)$ where (k), in this chapter, is confined to *integer* values, the first $N/2 - 1$ are a collection of positive-frequency phasors. It is sometimes sufficient to work with just this information. Another approach is usually more desirable and is easy to accomplish. Figure 2-1 illustrates a problem that occurs frequently when we use only one-half (first or second) of the phasors of an $X(k)$ sequence. Part (a) is a sine wave. Part (b) is its two-sided phasor spectrum using the DFT. Next (c) is the attempt to reconstruct the sine wave using only one-half (positive or negative) of the spectrum. The result is two sine waves in phase quadrature and half-amplitude values. Part (d) is the correct restoration using the two-sided phasor spectrum.

Incorrect usage of sequences can lead to mysterious difficulties, especially in more complicated situations, that can be difficult to unravel. We can also see that in part (c), where only the first (positive) half of the spectrum was used for reconstruction, the real part has the correct waveform but the wrong amplitude (in this case, 0.5), which may not be important. But this idea must be used carefully because it is often not reliable and can lead to false conclusions (a common problem). Also, the average power

Discrete-Signal Analysis and Design, By William E. Sabin
Copyright © 2008 John Wiley & Sons, Inc.

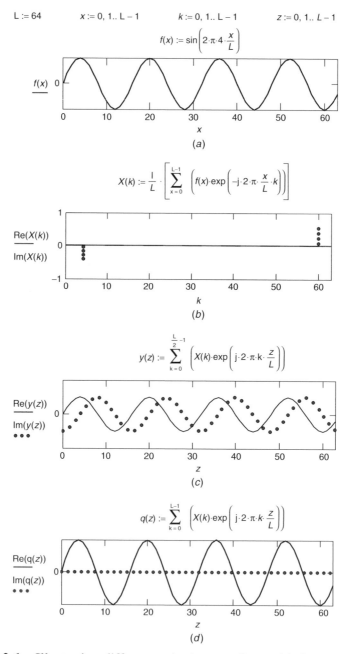

Figure 2-1   Illustrating differences in the use of one-sided sequences and two-sided sequences.

in Fig. 2-1 is zero because the two waves are $90°$ apart, which is not true for the actual sine wave.

ONE-SIDED SEQUENCES

Since an $X(k)$ *phasor* can be complex (Chapter 1), with a real (Re) part and an imaginary (jIm) part, and in practical electronic design we are usually interested in positive-frequency rather than two-sided, it is easy to convert a two-sided phasor *sequence $X(k)$* to a sum of positive-frequency sine and cosine time-domain *signals,* each at some amplitude and phase angle.

All of the samples of $x(n)$ are used to find each phasor $X(k)$ using the DFT of Eq. (1-2). A pair of complex-conjugate phasors then creates a sine or cosine signal with some phase and amplitude at a positive (k) between 1 and $(N/2) - 1$. We can then write an equation in the time domain that uses the sum of these sine and cosine signals and their phase angles at the positive frequencies.

Also, the entire sequence of $X(k)$ phasors finds positive-time (first half) $x(n)$ and negative-time (second half) $x(n)$ values using the IDFT, as in Eq. (1-8).

TIME AND SPECTRUM TRANSFORMATIONS

The basic idea is to use the DFT to transform a two-sided *time* sequence (Fig. 1-1d) into a two-sided *phasor* sequence. We then use a pair of these phasors, one from a positive-side location (for example, $k = 3$) and the other from the corresponding negative-side location ($k = N - 3$), to define a positive-frequency sine or cosine *signal* (at $k = 3$) at phase angle $θ(3)$. The pairs are taken from two regions of equal length, 1 to $(N/2) - 1$ and $(N/2) + 1$ to $N - 1$.

If the frequency phasors are known, the IDFT [Eq. (1-8)] returns the correct two-sided time sequence $x(n)$. We want the one-sided signal sine–cosine–θ spectrum that is derived from the two-sided phasor spectrum. To get the correct two-sided phasor spectrum from $x(n)$, we need the entire two-sided $x(n)$ sequence.

There are two types of phasor $X(k)$ sequences: those that have even symmetry about $N/2$ and those that have odd symmetry about $N/2$. Figure 2-2c, d, g, and h have odd symmetry. In either case we can add the *magnitudes* of the two phasors, which tells us that we have a true signal of some kind if both phasors are greater than zero. We then find the frequency, amplitude, and phase of each phasor. Mathcad then looks at each pair of phasors and determines their real and imaginary parts and calculates the sine- or cosine-wave time sequence, its amplitude, and its phase angle. Several different frequency components of a single wave can be determined in this manner.

Equations (1-5) to (1-7) showed how to organize the two sides of the (k) spectrum to get pairs. A simple Mathcad Program (subroutine) performs these operations automatically. Figures 2-3, 2-4, 4-1, 6-4, 6-5, 8-1, 8-2, and 8-3 illustrate the methods for a few of these very useful little "programs" that are difficult to implement without their logical and Boolean functions. The Mathcad Help (F1) "programming" informs us about these procedures.

The special frequency $k = N/2$ is not a member of a complex-conjugate pair and delivers only a real or complex number which is not used here but should not be discarded because it may contain significant power. We recall also that $k = 1$ is the fundamental frequency of the wave and (k) values up to $(N/2) - 1$ are harmonics (integer multiples) of the $k = 1$ frequency, so we choose the value of N and the frequency scaling factors discussed in Chapter 1 to assure that we are using all of the important harmonics. We can then ignore $k = N/2$ with no significant loss of data. Verify by testing that N is large enough to assure that all important pairs are used. We can then prepare a sine–cosine–θ table, including dc bias $(k = 0)$ using Eq. (1-4) and perform graphics plots.

Figure 2-2 shows the various combinations of the two-sided spectrum for the eight possible types of sine and cosine, both positive or negative and real or imaginary. Real values are solid lines and imaginary values are dashed lines. For any complex spectrum, defined at integer values of (k), the complex time-domain sine and cosine amplitudes and their phase angles are constructed using these eight combinations. Mathcad evaluates the combinations in Fig. 2-2 and plots the correct waveforms.

The example in Eq. (2-1) is defined for our purposes as the sum of a real cosine wave at $k = 2$, amplitude 5, and a real sine wave at $k = 6$,

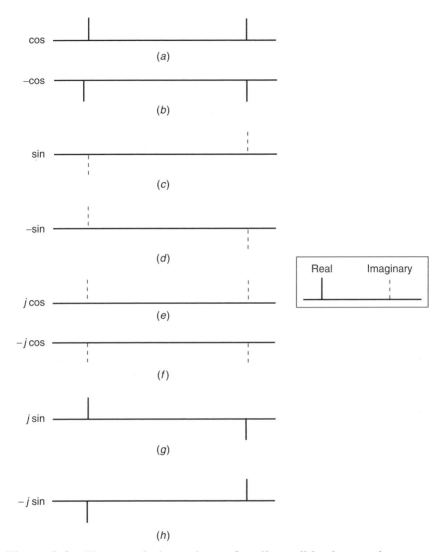

Figure 2-2   Phasor polarity and type for all possible sine–cosine waves. Solid lines: real; dashed lines: imaginary.

amplitude 8. The spectrum is positive-sided at 2 and 6.

$$x(n) = 5 \cos\left(2\pi\frac{n}{N}2\right) - j\,8\cos\left(2\pi\frac{n}{N}6\right) \qquad (2\text{-}1)$$

In Fig. 2-3 the (n) step size 0.1 from 0 to $N-1$ provides smooth curves in the real and imaginary $x(n)$ graphs. Figure 2-3 shows the following:

- The real cosine spectrum at $k=2$ and 14, amplitude $2.5+2.5= +5.0$ (see Fig. 2-2a).
- The $-j$ cosine spectrum at $k=6$ and 10, amplitude $-4\ -4=-8$ (see Fig. 2-2f).
- The phase at $k=2$ and 14 $(=0°)$.
- The phase at $k=6$ and 10 $(=-90°)$.

Observe that Mathcad provides the correct two-sided phasors (Fig. 2-2) that the subsequent two-sided IDFT restores to the input $x(n)$. If $x(n)$ is viewed from 0 to $N-1$, the positive frequencies 2 and 6 (number of cycles per record length) are visible.

In more complicated situations it is a good idea to avoid possible confusion by making sure that all of the $\mathrm{Re}[X(k)]$ and $\mathrm{Im}[X(k)]$ phasor pairs are combined into the correct one-sided real and imaginary sine and cosine positive-frequency constituents, as defined in Fig. 2-2. For the example in Fig. 2-3, the one-sided output is $+5$ cosine at $f=2$ at $0°$ and $-j8$ cosine at $f=6$ at $-90°$.

Note the use of the Mathcad function $[\mathrm{atan2}\ (\mathrm{Re}(x),\ \mathrm{Im}(x))]\cdot180/\pi°$, which covers the range $\pm180°$, as compared with $\mathrm{atan}(x)\cdot180/\pi°$, which only covers $\pm90°$.

There are some possibilities for "de-cluttering" the results:

- If $\mathrm{Im}(X(k))<0.0001$, set $\phi(k)=0$ to avoid clutter in the phase data.
- If $\mathrm{Re}(X(k))<0.0001$, set $\mathrm{Re}(X(k))=0.0001$.
- If $j\ \mathrm{Im}(X(k))>j1000$, set $j\mathrm{Im}(X(k))=j1000$.
- If $j\ \mathrm{Im}(X(k))<-j1000$, set $j\mathrm{Im}(X(k))=-j1000$.
- Scale the problem to avoid values of $X(k)\ <0.001$ or >1000.

In the next example we use just the cosine wave with a phase advance of $+60°=(\pi/3\ \text{radians})$:

$$X(k)=\frac{1}{N}\sum_{n=0}^{N-1}\left\{7\cos\left[2\pi\frac{n}{N}4+\left(60\cdot\frac{\pi}{180}\right)\right]\right\}\exp\left(-j2\pi\frac{n}{N}k\right)$$

$$(2\text{-}2)$$

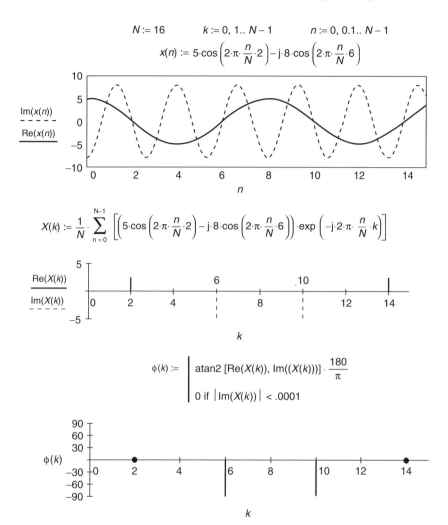

Figure 2-3 Example of two-tone signal and its spectrum components and phase (deg).

Figure 2-4 shows that a + 60° component has been added to the cosine component, as Fig. 2-2d confirms. The amplitude of the cosine segment is 1.75 + 1.75 = 3.5V. The amplitude of the − sin θ component is 3.03 + 3.03 = 6.06V. The phase is shown as + 60°.

In these two examples we started with the equations for the signal, then by examining the phasors in Fig. 2-2 we verified that they coincided with

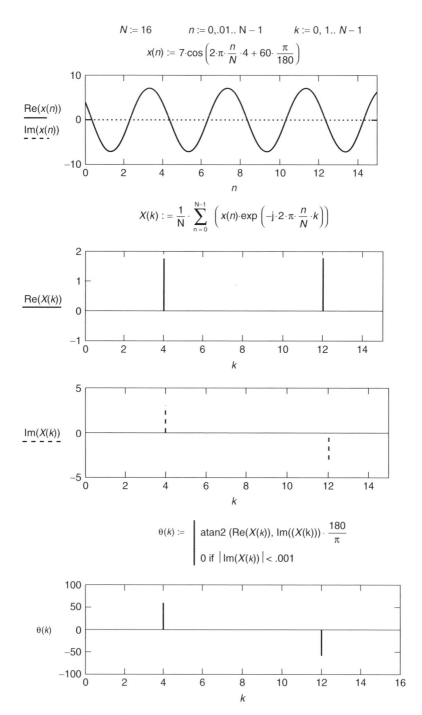

$$N := 16 \qquad n := 0, .01 .. \, N-1 \qquad k := 0, 1 .. \, N-1$$

$$x(n) := 7 \cdot \cos\left(2 \cdot \pi \cdot \frac{n}{N} \cdot 4 + 60 \cdot \frac{\pi}{180}\right)$$

$$X(k) := \frac{1}{N} \cdot \sum_{n=0}^{N-1} \left(x(n) \cdot \exp\left(-j \cdot 2 \cdot \pi \cdot \frac{n}{N} \cdot k\right)\right)$$

$$\theta(k) := \begin{vmatrix} \text{atan2 } (\text{Re}(X(k)), \text{Im}((X(k)))) \cdot \dfrac{180}{\pi} \\[2mm] 0 \text{ if } |\text{Im}(X(k))| < .001 \end{vmatrix}$$

Figure 2-4 Sine wave with 60 phase advance.

the given signal. In practice we might not know the waveform ahead of time and we would be asked to examine the phasor amplitudes and phases and use Fig. 2-2 to determine the signal.

We will find uses for the positive-side techniques described in this chapter and others. Because of the simplicity and the familiarity with positive-domain frequency usage in many practical engineering situations, the sine−cosine−θ approach described here is encouraged.

When setting up a problem to get the line spectrum, be sure that (k), the frequency index, is defined as an integer array. In Mathcad the assignment statement $k := 0, 1 \ldots N - 1$ works. Otherwise, the DFT generates a lot of complex answers for noninteger values of k and the spectrum becomes "smeared," which obscures the desired line spectrum answers. This general rule is a good idea in discrete wave analysis problems. Operate at integral values of (n) (time domain) and (k) (frequency domain) if possible. Procedures for dealing with noninteger values will be covered in later chapters, especially Chapters 3 and 4.

The calculations in Example 2-1 demonstrate how the two-sided DFT $X(k)$ of the two-sided $x(n)$, converted to positive-frequency $X(k)$ only, can be an excellent spectrum amplitude and phase analyzer for a nonlinear device, circuit, or system in the real-world positive frequency domain. Adjustments to signal level(s) of two or more signals and device biasing parameters can very quickly present a picture of the spectrum response of the circuit. This is an almost-free Spectrum Analyzer that can be very accurate and versatile over a very wide frequency range if (a big "if") the nonlinear devices used are defined correctly. The main requirement is that we have either an equation or an example data file for the device.

One interesting experiment is to look at nonlinear reactions to a two-tone signal, then vary the amplitude of a strong third signal that is on a third out-of-band frequency to see the degradation of the in-band two-tone signal. This can be very illuminating and very practical.

Example 2-1: Nonlinear Amplifier Distortion and Square Law Modulator

To get some hands-on experience, this example will look at the intermodulation (mixing) products of an amplifier circuit that is not perfectly linear. We will use the DFT [Eq. (1-2)] to get the spectrum and IDFT (Eq. 1-8) to return to the time domain with the confidence that these two equations, especially in discrete sequences, do not require linearity or superposition. We will use this idea frequently in this book.

A two-tone input signal of adjustable peak amplitude will be processed by a circuit that has a certain transfer characteristic which is similar to the Child-Langmuir equation [Seely, 1956, pp. 24 and 28, Eq. (2-14)] as derived in the early 1920s from Poisson's equation for the electric field in space-charge-limited diodes and also many common triode vacuum tubes:

$$I_{out}(n) \approx K V_g(n)^{1.5} \tag{2-3}$$

The input (base-to-emitter or grid-to-cathode) two-tone signal at frequencies f_1 and f_2 is

$$V_g(n) = V_s \left[\cos\left(2\pi \frac{n}{N} f_1\right) + \cos\left(2\pi \frac{n}{N} f_2\right) \right] + V_{dc} \tag{2-4}$$

V_s is the peak amplitude (1/2 of pk-to-pk) of each of the two input signals. V_{dc} is a bias voltage that determines the dc operating point for the particular device. This and a reasonable V_s value are found from Handbook V-I curves (the maximum peak-to-peak signal is four times V_s). The peak-to-peak ac signal should not drive the device into cutoff or saturation or into an excessively nonlinear region. Figure 2-5 is a typical approximate spectrum for the two-tone output signal. The input frequencies are f_1 and f_2, and the various intermodulation products are labeled. Adjusting V_s and V_{dc} for a constant value of peak *desired* per-tone output shows how distortion products vary. Note also the addition of 70 dB to the vertical axis. This brings up the levels of weak products so that they show prominently above the zero dB baseline (we are usually interested in the dB *differences* in the spectrum lines). Note also that the vertical scale for the spectrum values is the *magnitude* in dB because the actual values are in many cases complex, and we want the magnitude and not just the real part (we neglect for now the phase angles).

Note that in this example we let Mathcad calculate $V_{sig}(n)^{1.5}$ directly (the easy way), not by using discrete math (the hard way), just as we do with the exp($\cdot$), sin($\cdot$), cos($\cdot$) and the other functions. We are especially

$N := 64$ $\begin{array}{l} n := 0, 1.. 63 \\ k := 0, 1.. 63 \end{array}$ $Vin(n) := .25 \cdot n$

$Vdc := -8$ $Vsig(n) := Vdc + 1.5 \cdot \left(\cos\left(2 \cdot \pi \cdot \dfrac{n}{N} \cdot 4\right) + \cos\left(2 \cdot \pi \cdot \dfrac{n}{N} \cdot 5\right) \right)$

$Vout(n) := Vsig(n)^{1.5}$ $Fip(k) := \dfrac{1}{N} \cdot \displaystyle\sum_{n=0}^{N-1} \left(Vout(n) \cdot \exp\left(-j \cdot 2 \cdot \pi \cdot \dfrac{n}{N} \cdot k\right) \right)$

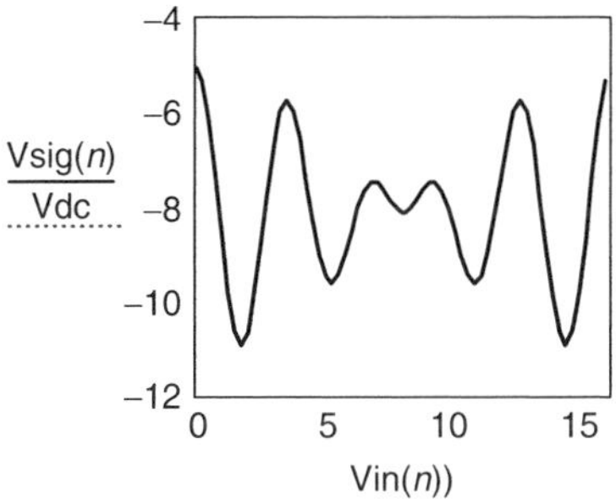

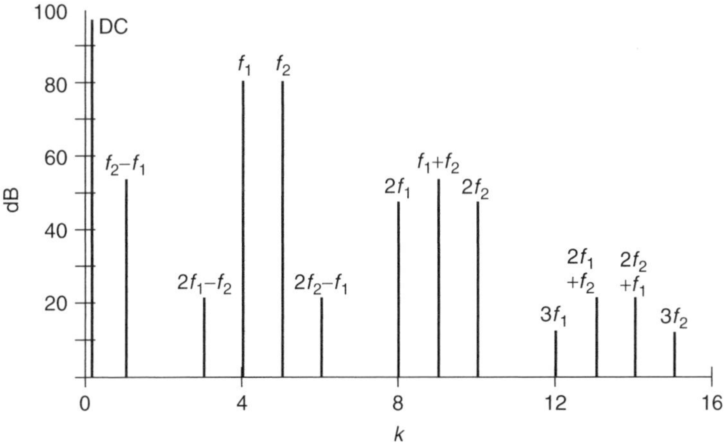

Figure 2-5   Intermodulation measurements on an amplifier circuit.

interested in discrete sequences and discrete ways to process them, but we also use Mathcad's numerical abilities when it is sensible to do so. In embedded signal-processing circuitry, machine language subroutines do all of this "grunt" work. In this book we let Mathcad do it in an elegant fashion.

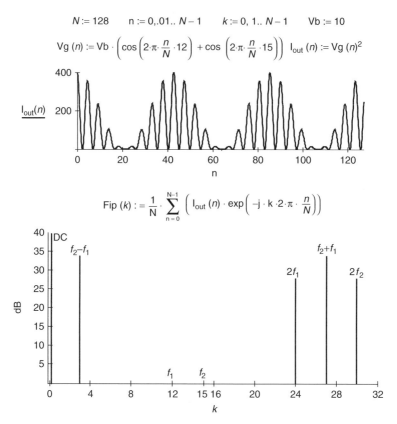

Figure 2-6 Square-law amplifier, mixer, and frequency doubler.

In Fig. 2-6 the exponent in Eq. (2-3) is changed from 1.5 to 2.0. This is the well-known *square-law device* that is widely used as a modulator (mixer) or frequency doubler [Terman, 1943, p. 565]. Note the absence of dc bias (optional). We see that frequencies f_1 and f_2 have disappeared from the output, the sum and difference of f_1 and f_2 are prominent and the second harmonics of f_1 and f_2 are strong also. Higher-order IMD products have also vanished.

The nonlinearity in Eq. (2-3) can be customized for a wide variety of devices, based on their transfer characteristics, to explore ac circuit performance. For example, Eq. (2-3) can be in the form of an N-point Handbook lookup table for transistor or tube V–I curves. Pick 16 equally spaced values of $V_{in}(n)$ for $n = 0$ to 15 and estimate as accurately as possible the corresponding values of $I_{out}(n)$. Then get the positive frequency spectrum for low-order (2nd or 3rd) intermodulation products. Nonlinear circuit simulation programs such as Multisim can explore these problems in greater detail, using the correct dc and RCL components and accurate slightly nonlinear device models.

Example 2-2: Analysis of the Ramp Function

This chapter concludes with an analysis of the "ramp" function in Fig. 2-7a. It is shown in many references such as [Zwillinger, 1996, p. 49]. Its Fourier series equation in the Reference is

$$f(x) = \frac{1}{2} - \sum_{k=1}^{\infty} \frac{1}{\pi k} \sin \frac{2\pi xk}{L} \tag{2-5}$$

where x is the distance along the x-axis. The term 1/2 is the average height of the ramp, and x always lies between 0 and $+L$. The sine-wave harmonics (k) extend from 1 to ∞, each with peak amplitude $1/\pi k$. For each value of (k), $f(x)$ creates (k) sine waves within the length L. The minus sign means that the sine waves are inverted with respect to the x axis.

In Fig. 2-7 the discrete form of the ramp is shown very simply in part (a) as $x(n) = n/N$ from $0 \le n \le N-1$. We then apply the DFT (Eq. 1-2) to get the two-sided phasor spectrum $X(k)$ from 0 to $N - 1$ in part (b). The following comments help to interpret part (b):

- The real part has the value $X(0) = 0.5$ at $k = 0$, the dc value of the ramp. The actual value shown is 0.484, not 0.5, because the value 0.5 is approached only when the number of samples is very large. For $N = 2^9 = 512$, $X(0)$ is 0.499. This is an example of the approximations in discrete signal processing. For very accurate answers that we probably will never need, we could use 2^{12} and the FFT.
- The real part of $X(k)$ from $k = 1$ to $N - 1$ is negative, and part (c) shows that the sum of these real parts is -0.484, the negative of $X(0)$. In part (c) the average of the real part from 0 to $N - 1$ is zero, which is correct for a spectrum of sine waves with no dc bias.

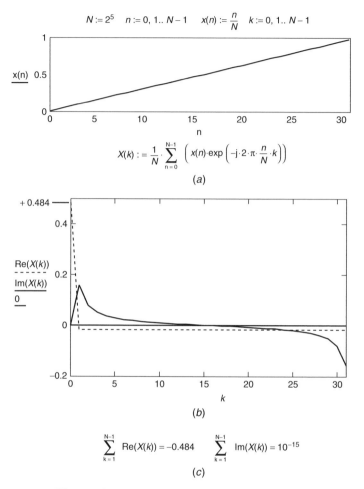

$$N := 2^5 \quad n := 0, 1 .. N-1 \quad x(n) := \frac{n}{N} \quad k := 0, 1 .. N-1$$

$$X(k) := \frac{1}{N} \cdot \sum_{n=0}^{N-1} \left(x(n) \cdot \exp\left(-j \cdot 2 \cdot \pi \cdot \frac{n}{N} \cdot k \right) \right)$$

(a)

(b)

$$\sum_{k=1}^{N-1} Re(X(k)) = -0.484 \qquad \sum_{k=1}^{N-1} Im(X(k)) = 10^{-15}$$

(c)

Figure 2-7   Analysis of a ramp function.

- The imaginary part of the plot is positive-going in the first half, negative-going in the second half, and zero at $N/2$. Referring to Fig. 2-2, this arrangement of polarity agrees with the $-\sin$ diagram, as it should.

- Applying the "Mathcad X-Y Trace" tool to the imaginary part of the spectrum plot in part (b), we find that the two sides are odd-symmetric (Hermitian) about $N/2$. For this reason, the relative phase is zero for these sine waves (they all begin and end at zero phase).

- To get the one-sided spectrum, we combine $k = 1$ with $k = N - 1$ and so on from $1 \leq k \leq (N/2 - 1)$. For $k = 1$ to 5, using the Mathcad X-Y Trace tool, we get 0.3173, 0.1571, 0.1030, 0.0754, and 0.0585. The Trace tool is a very useful asset.

REFERENCES

Seely, S., 1956, *Radio Electronics*, McGraw-Hill, New York. (Also Google, "Child-Langmuir.")

Terman, F.E., 1943, *Radio Engineer's Handbook,* McGraw-Hill, New York.

Zwillinger, D., ed., 1996, *CRC Standard Mathematical Tables and Formulae* 30th ed., CRC Press, Boca Raton, FL.

3

Spectral Leakage and Aliasing

SPECTRAL LEAKAGE

The topics in the title of this chapter are concerned with major difficulties that are encountered in discrete signal waveform analysis and design. We will discuss how they occur and how we can deal with them. The discussion still involves eternal, steady-state discrete signals.

Figure 3-1a shows the "leaky" spectrum of a complex phasor using the DFT [Eq. (1-2)] at $k = 7.0$ (Hz, kHz, MHz, or just 7.0) whose input signal frequency (k) may be different than 7.0 by the very small fractional offset $|\varepsilon|$ shown on the diagram. For $|\varepsilon|$ values of 10^{-15} to 10^{-6}, the spectrum is essentially a "pure" tone for most practical purposes. The dots in Fig. 3-1a represent the maximum spectrum attenuation at *integer* values of (k) for each of the offsets indicated. A signal at 7.0 with the offset indicated produces these outputs at the other *exact* integer (k) values. Figure 3-1b repeats the example with $k = 15.0$, and the discrete spectrum

Discrete-Signal Analysis and Design, By William E. Sabin
Copyright © 2008 John Wiley & Sons, Inc.

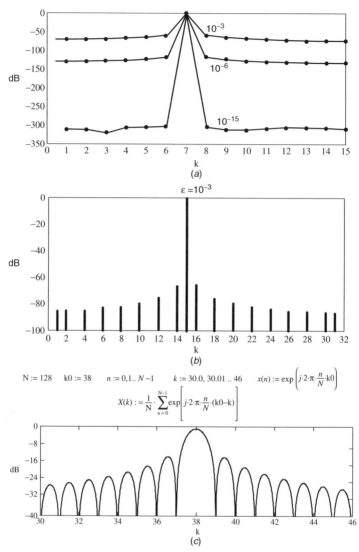

Figure 3-1 (a) Spectrum errors due to fractional error in frequency specification. (b) Line spectrum errors due to 10^{-3} fractional error in frequency. (c) Continuous spectrum at 38 and at minor spectral leakage loops. (d) Continuous spectrum of real, imaginary, and magnitude over a narrow frequency range. (e) Time domain plot of the $u(t)$ signal, real sine wave plus dc component. Imaginary part of sequence $= 0$. (f) Real and imaginary components of the spectrum $U(k)$ of time domain signal $u(t)$. (g) Incorrect way to reconstruct a sine wave that uses fractional values of time and frequency increments.

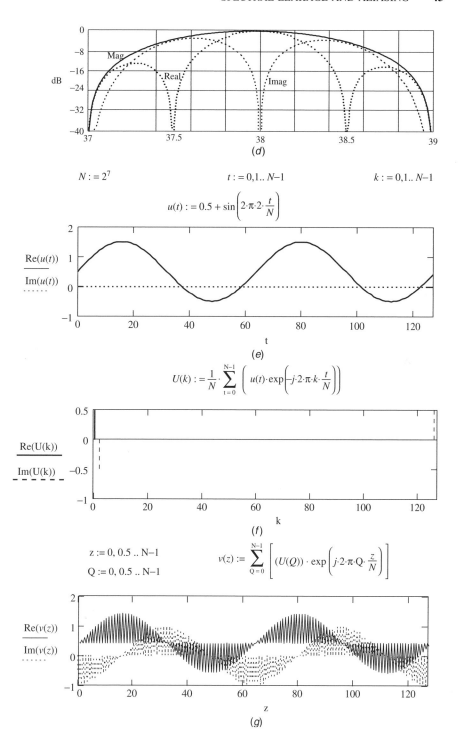

$$N := 2^7 \qquad\qquad t := 0, 1 .. N-1 \qquad\qquad k := 0, 1 .. N-1$$

$$u(t) := 0.5 + \sin\left(2 \cdot \pi \cdot 2 \cdot \frac{t}{N}\right)$$

(e)

$$U(k) := \frac{1}{N} \cdot \sum_{t=0}^{N-1} \left(u(t) \cdot \exp\left(-j \cdot 2 \cdot \pi \cdot k \cdot \frac{t}{N}\right) \right)$$

(f)

$$z := 0, 0.5 .. N-1 \qquad\qquad v(z) := \sum_{Q=0}^{N-1} \left[(U(Q)) \cdot \exp\left(j \cdot 2 \cdot \pi \cdot Q \cdot \frac{z}{N}\right) \right]$$

$$Q := 0, 0.5 .. N-1$$

(g)

Figure 3-1 (*continued*)

lines are drawn for an $|\varepsilon|$ value of 10^{-3}. Again, the spectrum of the signal is shown at integer values of (k). If we are able to confine our interest to these integer values of (k), these figures characterize the performance of the DFT for an input that is very close in frequency to an integer (k) value. Exact values of (k) give optimum frequency resolution between adjacent values of (k), which is why they are preferred when possible. It is not always possible, as we will see in Chapter 4.

Figure 3-1a and b do not tell the entire story. Assume the following $x(n)$ complex input voltage *sequence* at frequency $k_0 = 38.0$ in Eq. (3-1)

$$x(n) = \exp\left(j2\pi \; \frac{n}{N} \; k_0 \right), \quad k_0 = 38.0 \qquad (3-1)$$

Using Eq. (1-2) for the DFT, $N = 128$, and $k = 30$ to 46 in steps of 0.01, the phasor frequency response $X(k)$ is (review p. 24)

$$X(k) = \frac{1}{N} \sum_{n=0}^{N-1} x(n) \exp\left(-j2\pi \frac{n}{N} k \right)$$

$$= \frac{1}{N} \sum_{n=0}^{N-1} \exp\left(j2\pi \frac{n}{N} k_0 \right) \exp\left(-j2\pi \frac{n}{N} k \right)$$

$$= \frac{1}{N} \sum_{n=0}^{N-1} \exp\left[j2\pi \frac{n}{N} (k_0 - k) \right] \qquad (3-2)$$

Mathcad finds the real part, the imaginary part and the magnitude of the complex exponential (phasor) at each non integer value of (k).

Figure 3-1c shows the magnitude in dB on a 0 to -40 dB scale. This is a "selectivity" curve (ratio in dB below the peak) for the DFT. At 37.5 or 38.5, for example, the response is down 3.92 dB. An input signal at either of these frequencies will show a reduced output at 38.0 (the *scalloping effect*). At any other input signal frequency k_0 that lies between adjacent integer-k values, we can repeat Eq. (3-2) to find the spectrum for that k_0, and we suggest that the reader experiment with this for additional insight.

The last term in Eq. (3-2) is a virtual *scalar* spectrum analyzer. At each 0.01 increment of frequency, it calculates and then sums, for each of 128 time values in Fig. 3-1c, the *magnitude* of $X(k)$ in dB below the reference level (0 dB). Note that the loop width at 38 is 2.0 and other loop widths are 1.0, but the response is zero at 37 and 39. If we calculate the real and imaginary parts of the spectrum, we could have a *vector* network analyzer. In other words, the Mathcad Worksheet can be taught how to measure forward and reflected complex waves by using the mathematical equivalent of an ideal wideband four-port directional coupler [Sabin, 1995, p. 3]. S-parameters can be derived from these complex waves.

Figure 3-1d is a close-up of the spectrum between 37 and 39, showing the magnitude, real part, and imaginary part, all in dB. It is obvious that the spectrum becomes very complex, although the magnitude decreases smoothly. At 38.5 the phase is $-90°$, and at 37.5 the phase is $+90°$, (the graph shows only magnitudes in dB). At 37.0, 38.0 and 39.0 the phase is $0.0°$. The magnitude plot is the same as that in Fig. 3-1c over the same frequency range.

Further mathematical discussion of these waveforms involves the sinc function (see, for example, [Carlson, 1986, p. 25]).

Methods of reducing the leakage by *windowing* will be covered in Chapter 4, but it is always desirable, when feasible, to stay as close as possible to integer values of (n) and (k). This can often be arranged using the scaling methods for time and frequency described in Chapter 1. For example, let each (k) represent a smaller signal frequency band, say 1 kHz instead of 10 kHz, and reduce the frequency sweep range by a factor of 10. Also reduce the amplitude scale. This is what a spectrum analyzer does, and Fig. 3-1d is an example of how this works (see Fig. 3-2).

One common problem with leakage is that it can obscure spectrum amplitudes on closely adjacent frequencies, making frequency resolution uncertain. For example, in Fig. 3-1c and d the very small slope at the spectrum peak (38.0) makes it difficult to distinguish closely adjacent frequencies. Zooming in on the peak, both horizontally and vertically, can be helpful but there is always some uncertainty about the true frequency. A familiar example of this problem is the ordinary digital frequency counter. No matter how many digits are displayed, the uncertainty is at least $\pm\frac{1}{2}$ the least significant digit. The scaling procedure is a good way to improve this problem.

At this point we bring up another example, similar to Fig. 2-1, of a common problem that occurs in sequence analysis. Figure 3-1e shows a sine wave with an added dc bias. The DFT (see Fig. 3-1f) shows the correct line spectrum. When we try to reconstruct the sine wave using fractional instead of integer values of time and frequency in order to improve the resolution the result is not very good (see Fig. 3-1g). The second attempt, using integer values of time $x(n)$ and frequency $X(k)$, is successful (see Fig. 3-1e).

When we design a problem on the computer we usually have the option of *specifying* the exact frequency to within a very small error as shown in Fig. 3-1a. If the data is from a less exact source, we may be able to *assign* an exact frequency that is equal to 2^M ($M =$ integer) using the scaling methods in Chapter 1. If the computer program is trying to *determine* the true frequency, the problem is the same as the frequency counter problem. However, phase noise (related to frequency jitter) and other random or pseudo random problems can often be improved by using statistical methods such as *record averaging*, which will be described in later chapters. It is also possible to insert experimentally a small frequency offset into the problem that puts a deep notch at some integer frequency, as shown in the following Example (3-1).

Finally, Chapter 4 discusses the subject of *windowing*, which can relax somewhat the requirements for close alignment with integer values of (n) and (k).

Example 3-1: Frequency Scaling to Reduce Leakage

Frequency scaling can improve spectral leakage. Figure 3-2 illustrates how two signals on adjacent frequencies (24 and 25) interact when their frequency values are not aligned with integer-valued frequencies. Two values of percentage frequency offset are shown, 1.0% (24.24 and 25.25) and 0.1% (24.024 and 25.025). The improvement in dB isolation between the two can be significant in many applications.

ALIASING IN THE FREQUENCY DOMAIN

This is another important subject that occurs very often and requires careful attention. Figure 3-3 shows the amplitude of a discrete *two-sided complex phasor spectrum* that is centered at zero with frequency range -16 to $+16$. For a steady-state infinite sequence, duplicate two-sided spectra

$$X(k) := \frac{1}{N} \cdot \sum_{n=0}^{N-1} \exp\left[j \cdot 2 \cdot \pi \cdot \frac{n}{N} \cdot (k0 - k)\right] \qquad XX(k) := 20 \cdot \log(|X(k)|)$$

$$Z(k) := \frac{1}{N} \cdot \sum_{n=0}^{N-1} \exp\left[j \cdot 2 \cdot \pi \cdot \frac{n}{N} \cdot (k1 - k)\right] \qquad ZZ(k) := 20 \cdot \log(|Z(k)|)$$

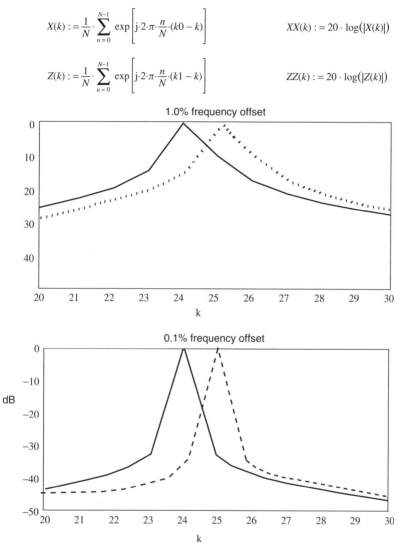

Figure 3-2 Spectral leakage versus frequency offset.

are centered at, ± 32, ± 64, and so on. We see that the two phasor spectra overlap considerably at ± 16, and also at 0 and ± 32 to a lesser extent.

Figure 3-3 is a set of discrete-frequency Fourier transformations of an eternal steady-state discrete-time process. Each line is the amplitude $X(k)$ of a rotating phasor at frequency (k) (review Fig. 1-5). The discrete phasor

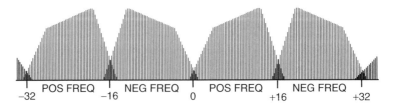

POS FREQ   NEG FREQ   POS FREQ   NEG FREQ
−32   −16   0   +16   +32

Figure 3-3   Phasor aliasing in the frequency domain.

spectrum is a collection of these individual rotating phasors. The value at each (k) is determined by the $\Sigma(0 \leq n \leq N - 1)$ indicated in Eq. (1-2). Also indicated at $k = 0$ is a dc value.

There are two common explanations for this overlap at $k = 0$. One is that they are created by conditions that exist in the $x(n)$ time-domain sequence for values of $n < 0$. This is time-domain aliasing, which is discussed at greater length later in this chapter. These values of $x(n)$ can often transform into values of $X(k)$ that are in the *phasor* negative-frequency range. For example, a capacitor or inductor that has a changing energy storage before $n = 0$ can lead to this overlap from $k < 0$ to $k = 0$ to $k > 0$. This situation also replicates at ± 32, ± 64 and so on. These $(n < 0)$ conditions are parts of the eternal $x(n)$ sequence that repeats over and over in time and therefore also in frequency. If the value of $X(k)$ at $k = 0$ is not zero, there is a permanent dc component in the spectrum. The entire patterns of $x(n)$ and $X(k)$ repeat endlessly.

The second explanation is considered next. When a phasor spectrum is created by frequency translation from radio (RF) or intermediate (IF) frequencies to baseband, phasors that overlap into negative frequencies can be created. We will learn how to deal with this problem.

The overlap condition illustrated in Fig. 3-3 is *aliasing*, meaning that parts of one spectrum segment invade its two neighbors and become associated, often inextricably, with them. At 0, ± 16, and ± 32, etc., a negative-frequency phasor region collides with a positive-frequency phasor region. Note that the spectra in Fig. 3-3 are symmetrical with respect to 0, ± 16, and ± 32, but very often they are not symmetrical, as illustrated in Fig. 1-3. The overlap regions produce interactions between the two spectra that can be very difficult to deal with and to separate. Because the phase relationships in the overlapping negative- and positive-frequency regions may not be well known, it is usually difficult to predict the exact behavior in the alias region. We will look at ways to deal with these overlaps so that aliasing is reduced, if not to zero, then at least to sufficiently low levels that it becomes unimportant. In many cases a small amount of aliasing can be tolerated.

We emphasize that Fig. 3-3 deals with *phasors*. Mathematically, physical sine and cosine waves as defined explicitly in Eq. (2-1) do not exist as *separate* entities at negative frequencies, despite occasional rumors to the contrary. (Instruments such as spectrum analyzers, oscilloscopes, etc. can be used to create certain visual illusions; the Wells-Fargo stage coach wheels in old-time Westerns often appear to be turning in the wrong direction.) For example, as compared to phasors that can rotate at positive or negative angular frequencies,

$$\cos(-\omega t) = \cos(+\omega t) \tag{3-3}$$

$$\sin(-\omega t) = -\sin(+\omega t) = \sin(+\omega t) \angle \pm 180° \tag{3-4}$$

We note also that the average *power* in any single phasor of any constant amplitude is zero $\pm$ the resolution of the computer. So the phasor at frequency (k) must never be thought of as a true signal that can light a light bulb or communicate.

The sine or cosine wave requires two phasors, one at positive frequency and one at negative frequency, and the result is at positive frequency. As an electrical signal the individual phasor is a mathematical concept only and not a physical entity (we often lose sight of this). Therefore, aliasing for sine and cosine spectra requires special attention, which we consider in this chapter.

Adjacent segments of the eternal steady-state positive-frequency sine–cosine spectrum can overlap at frequencies greater than zero, and it is a common problem. In Fig. 3-4a the positive-frequency eternal steady-state spectrum is centered at 5, 15, 25, 35, 45, and so on. Each side of each spectrum segment collides with an adjacent segment, producing alias regions. This spectrum pattern repeats every 10 frequency units. The plots are shown in Fig. 3-1 as continuous lines rather than discrete lines. We do this often for convenience, with the understanding that a discrete spectrum is what is intended.

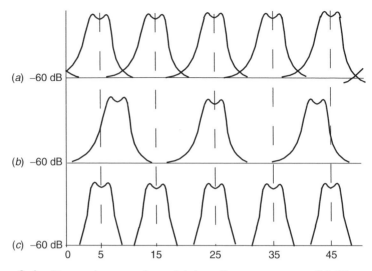

Figure 3-4   Removing overlaps (a) in adjacent spectra. (b) The separations of the BPF center frequencies are increased. (c) The BPF transition regions are made steeper. The 3-dB bandwidth is held constant.

Any one of these frequency patterns, for example, at 45, can be demodulated (mixed or converted) to baseband, and the original message, 0 to approximately 10, is recovered completely (if it is sampled adequately), along with any aliasing "baggage" that exists. The plot in Fig. 3-4b is similar to what the uncorrupted baseband spectrum might look like originally.

The baseline for Fig. 3-4 is 60 dB below the peak value. This choice represents good but not state-of-the-art performance in most communications equipment design. The desired RF band and the baseband can both be amplified. In communications equipment design this is a very common practice that produces the optimum *gain and noise figure distribution* in the signal path. In this regard, digital signal processors (DSP) require careful attention to signal and noise levels. The least few significant bits need to be functioning for weak signals or noise (noise dithered) and the maximum levels must not be exceeded *too* often.

The frequency conversion can also be, and very often is, upward from baseband to 45, etc. in a transmitter, and gain distribution is important there also.

The level of aliasing that is permitted is an important specification. For many typical radio systems this number can be between 30 dB (medium performance) and 60 dB (high performance), as shown in Fig. 3-4b and 3-4c. One satisfactory approach is shown in Fig. 3-4b, where the passband center frequencies are increased sufficiently. In some systems this method may have some economic or technical problems; it can sometimes be difficult and/or expensive to increase the 5 frequency (Fig. 3-4b) sufficiently. Improving the bandpass filtering of the signal before or after modulation (transmitter) or before or after demodulation (receiver) using improved filters with a constant value of passband width is very common and very desirable. This excellent approach is illustrated in Fig. 3-4c, where the 60 dB-bandwidth is reduced by appropriate filter design improvements. The distance at -60 dB between the frequency segments is a "guard band", which makes it easier to isolate any segment using a practical bandpass filter.

The plots of Fig. 3-4 can also be a set of *independent* signals whose spectra overlap. This is a very important design consideration. The result is not aliasing, but adjacent channel crosstalk (interference). The methods of reducing this interference are the same as those that we use to reduce aliasing.

We often (usually) prefer to think in terms of positive-frequency sine–cosine waves as described in Chapter 2. This gives us an additional "classical" insight into aliasing. The situation is shown in Fig. 3-5a, where a spectrum centered at 600 kHz is translated to baseband by a local oscillator (L.O.). Because the lower frequencies of the signal spectrum appear to be less than zero, the output signal "bounces" off the zero-frequency axis and reflects upward in frequency. That is, the output frequency of a down-converting mixer is the *magnitude* of the difference between the input L.O. and the RF. Also implied are a coupling dc block and an internal short circuit to ground at zero frequency. Figure 3-5b shows the output of the mixer, including a zero-frequency notch, for each side of the input spectrum. The alias segment is moved to a high-frequency where it can interfere, perhaps nonlinearly, with the high-frequency region of the other part of the output. Additional improvement in the filter design at low frequencies is often needed to reduce this aliasing problem sufficiently. In Fig. 3-5b the two components of the output cannot be combined mathematically unless the nature of the signal and the phase properties of the filter over the entire frequency range are well known.

This example reminds us that some problems are better approached in the frequency domain. A time-domain analysis of a typical bandpass

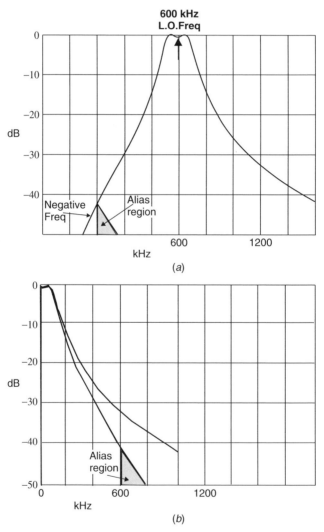

Figure 3-5   After conversion to baseband, the effect of aliasing is visible after being moved to a high frequency.

filter would answer some important questions. Imagine an input signal whose frequency changes too quickly as it traverses the frequency range of the filter, producing a spectrum distortion that is often seen in a spectrum analyzer. Time-domain analysis, followed by time-to-frequency conversion, might not inform us as easily regarding steady-state frequency aliasing.

An important task is to find the power that resides in the overlap (aliasing) region. We can do this by integrating the signal that resides in this region, whose spectral content is known (or estimated). The design goal is to assure that the power, usually the peak envelope power (PEP) that leaks into the alias (adjacent channel) region is an acceptable number of dB below the PEP that resides in the "desired" region. For radio communications equipment, 60 dB (commercial) or 50 dB (amateur) are readily achievable and quite satisfactory numbers. Some degradations of these numbers in transmitters and receivers are due to imperfections in circuit or system design: for example in a transmitter final amplifier and in a receiver RF "front-end". Example 3-2 illustrates this problem.

Example 3-2: Analysis of Frequency-Domain Aliasing

Figure 3-6 is a simplified example of (a) aliasing between two channels of a repetitive bandpass filter for a repetitive signal spectrum or (b) adjacent channel interference between two independent adjacent channel signals. The input signal for this example is Gaussian white noise with a spectral density of 0 dBm (1.0 mW) in a 1.0-Hz band, which is 10.0 W in the 10-kHz passband. The scanning instrument has a noise bandwidth of 300 Hz, so the passband noise power in that bandwidth is

$$\text{watts in } 300\,\text{Hz} = 10.0 \left(\frac{300}{10,000} \right) = 0.3\,\text{W} = 24.77\,\text{dBm}$$

$$\text{where dBm} = 10 \log \frac{0.3}{0.001} \tag{3-5}$$

The filter response is rectangular down to the -40, dBm (in a 1-Hz bandwidth) level and then slopes according to the dB-scale straight line over the region 0 to 6000 Hz between adjacent channels. In a 1.0-Hz bandwidth,

$$P(f)_{1\text{Hz BW}} \text{ dBm} = -40 \text{ dBm} - 20\left(\frac{f}{6000}\right) \text{ dB} \qquad (3\text{-}6)$$

We convert dBm per Hz to watts per Hz, integrate from 0 to 6000 Hz, and adjust to the 300-Hz instrument bandwidth:

$$P_{\text{out of band}}(\text{W}) = \frac{300}{6000} \int_0^{6000} P(f)\,df = 6.45 \times 10^{-6} \text{ W in 300 Hz}$$

$$(3\text{-}7)$$

The average $P_{\text{out of band}}$ in dBm in a 300-Hz band is -21.90 dBm. The ratio of in-band to out-of-band power for a 300-Hz bandwidth is $24.77 \text{ dBm} - (-21.90 \text{ dBm}) = 46.7$ dB. Note the method of employing dBm and dB in an equation. If both sides of the filter are considered (which they are not at this time), the ratio becomes 3 dB smaller.

We want the power in the alias zone shown in Fig. 3-6. To get this, integrate the out-of-band spectrum $P(f)$ from 3000 to 6000 Hz, multiply by 2 to get both halves of the alias zone, and adjust for the 300-Hz instrument bandwidth:

$$P_{\text{alias}} = 2\left(\frac{300}{3000}\right) \int_{3000}^{6000} P(f)\,df$$

$$= 2.345 \times 10^{-6} \text{ W} = -26.3 \text{ dBm}$$

$$(3\text{-}8)$$

The ratio of in-band power to alias band power between the two bands shown is $24.77 \text{ dBm} - (-26.3 \text{ dBm}) = 51.07$ dB. Subtract 3 dB for an additional alias band on the left side of the diagram.

The instrument for this measurement can be a calibrated spectrum analyzer. Since the noise signal is the same kind at every frequency and amplitude of interest in this example, we do not worry about the fact that the amplitude reading for noise on a spectrum analyzer is not quite the same as for a sine wave. The relative dB readings are correct. Also, in Eqs. (3-7) and (3-8) we are finding the *average* power over the specified band and then normalizing that average power to a

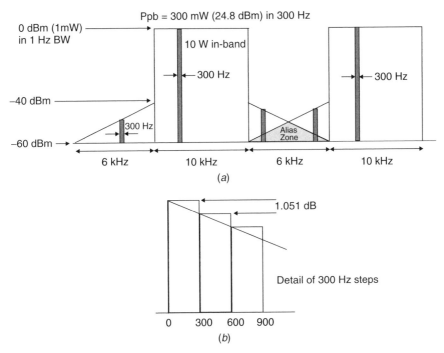

Figure 3-6 Power in the aliasing zone.

300 Hz bandwidth. It is important to not change any analyzer settings that might affect the internal adjustments or calibrations of the instrument. Finally, if the spectrum analyzer contains a high-quality tracking generator, it can be used as a sine-wave signal source instead of a noise generator. Figure 3-7 is the Mathcad worksheet used to perform the calculations for this example which could be a template reference for the procedure.

In more complicated (irregular) examples it may be necessary to divide the various frequency ranges into narrow non-overlapping frequency strips, to analyze each strip individually, and to combine the results in a manner similar to that suggested here. In wideband measurements it is often necessary to verify the instrument calibrations across the measurement frequency range and to eliminate spurious system responses that can invalidate the results.

$$Ppb := \frac{300}{10000} \times \int_{0}^{10000} .001 df \quad Ppb = 0.3 \quad \boxed{\text{watts in 300 Hz pass band}}$$

$$PdBm := 10 \times \log\left(\frac{Ppb}{.001}\right) \quad PdBm = 24.7712 \quad \boxed{\text{dBm in 300 Hz pass band}}$$

$$f := 0,1.. \ 6000 \quad \boxed{\text{out-of-band frequency index}}$$

$$dBm(f) := -40 - 20 \times \left(\frac{f}{6000}\right) \quad P(f) := 0.001 \times 10^{\frac{dBm(f)}{10}}$$

$$Poutb := \frac{300}{6000} \times \int_{0}^{6000} P(f) \, df$$

$$Poutb = 6.4493 \times 10^{-6} \quad PoutdBm := 10 \times \log\left(\frac{Poutb}{.001}\right) \quad PoutdBm = -21.9049$$

$$10 \times \log\left(\frac{2 \times Poutb}{Ppb}\right) = -44 \quad \boxed{\text{dB for sum of both out-of-band regions}}$$

$$Pazone := \frac{2 \times 300}{3000} \times \int_{3000}^{6000} P(f) \, df \quad \boxed{\text{watts in alias zone in 300 Hz band}}$$

$$Pazone := 2.345 \times 10^{-6} \quad \boxed{\text{watts in alias zone in 300 Hz band}}$$

$$10 \times \log\left(\frac{Pazone}{Ppb}\right) = -51.1 \quad \boxed{\text{dB below passband}}$$

Figure 3-7   Calculation of power in the alias zone of Fig. 3-6.

ALIASING IN THE TIME DOMAIN

Aliasing has been considered primarily in the $X(k)$ frequency domain, where bandlimited spectra overlap. But aliasing also occurs in the time domain, where periodic $x(n)$ *time* sequences similar in appearance to Figs. 3-3 and 3-4 overlap or are truncated or interrupted prematurely before the sample values become insignificant. An oscilloscope can easily show the overlap between two separate and independent time-sequence generators that are triggered alternately; the two could be triangular waves. After the DFT transformation the result is often an unacceptable modification of the spectrum. It is important that all of the significant data in the time-domain data record be obtained and utilized and that this record has sufficient resolution to include both high-frequency and low-frequency elements. Some smoothing or windowing of the time-domain waveform prior to the Fourier transformation may be desirable to reduce spurious high-frequency irregularities that might mask important results. These subjects are described in greater detail in Chapter 4.

REFERENCES

Sabin, W. E., 1995, The lumped element directional coupler *QEX* (ARRL), March.

Carlson, A. Bruce, 1986, *Communication Systems*, 3rd ed., McGraw-Hill, New York.

4

Smoothing and Windowing

In this chapter we consider ways to improve discrete sequences, including the reduction of data contamination and the improvement of certain time and frequency properties. Smoothing and windowing are useful tools for signal waveform processing in both domains. Simplifications with limited goals will be a desirable approach. The References in this chapter can be consulted for more advanced studies. Our suggested approaches are quite useful, where great sophistication is not required, for the processing of many commonly occurring discrete-signal waveforms.

SMOOTHING

Consider Fig. 4-1a and the rectangular discrete sequence $W(i)$ for (i) from 0 to 63 with amplitude 1.0 and drawn in continuous form for visual clarity. The simple Mathcad Program shown creates this sequence. Observe that the rectangle is delayed at the beginning and terminated early ("pedestal" would be a good name). This sequence is a particular kind of simple *window* that is useful in many situations because of the zero-value segments

Discrete-Signal Analysis and Design, By William E. Sabin
Copyright © 2008 John Wiley & Sons, Inc.

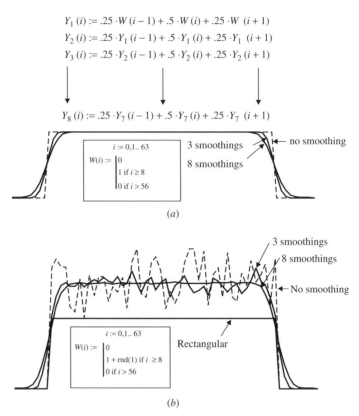

$$Y_1(i) := .25 \cdot W(i-1) + .5 \cdot W(i) + .25 \cdot W(i+1)$$
$$Y_2(i) := .25 \cdot Y_1(i-1) + .5 \cdot Y_1(i) + .25 \cdot Y_1(i+1)$$
$$Y_3(i) := .25 \cdot Y_2(i-1) + .5 \cdot Y_2(i) + .25 \cdot Y_2(i+1)$$

$$Y_8(i) := .25 \cdot Y_7(i-1) + .5 \cdot Y_7(i) + .25 \cdot Y_7(i+1)$$

$i := 0,1.. 63$

$W(i) := \begin{vmatrix} 0 \\ 1 \text{ if } i \geq 8 \\ 0 \text{ if } i > 56 \end{vmatrix}$

3 smoothings

8 smoothings

no smoothing

(a)

3 smoothings

8 smoothings

No smoothing

$i := 0,1.. 63$

$W(i) := \begin{vmatrix} 0 \\ 1 + \text{rnd}(1) \text{ if } i \geq 8 \\ 0 \text{ if } i > 56 \end{vmatrix}$

Rectangular

(b)

Figure 4-1 Smoothing operation on a discrete signal waveform: (a) without added random noise; (b) with added random noise.

at each end that are known as guardbands that greatly reduce spillover (aliasing, Chapter 3) into adjacent regions.

Figure 4-1 then shows how, at each value of (i), the three-point *smoothing sequence* 0.25, 0.5, 0.25 is applied to the data points $W(i-1)$, $W(i)$, and $W(i+1)$, respectively, to get $Y_1(i)$. We then repeat the operation, using the $Y_1(i)$ values to get the $Y_2(i)$ values, then the $Y_2(i)$ values to get the $Y_3(i)$ values, and so on. We pause at Y_3 to view the intermediate results, and see that the edges of the rectangle have been "softened" and the pattern is extended away from the boundaries of the rectangle. We then continue on to Y_8, and the pattern becomes pretty well stabilized. In other words, where the Y values become small, the smoothing sequences produce less and less effect, so that further spreading becomes "negligible". Observe also that the maximum amplitude of the smoothed sequence is essentially 1.0. The choice of smoothing sequence values with the sum $\Sigma[0.25 + 0.5 + 0.25] = 1.0$ is responsible for this result.

In Fig. 4-1b some noise, the random number function rnd(1.0) with values from (0 to +1), has been added to the rectangular region as shown in the Mathcad program. After eight smoothing operations the noise has been greatly reduced and the maximum amplitude is nearly the 1.0 of the rectangle plus the average of the noise component, which is 0.5.

The value of root-mean-square (RMS) noise and the ratio in dB of RMS noise to the highly smoothed signal within the rectangular region are matters of practical interest. The procedure is to:

1. Find the partially smoothed sequence $Y_3(i)$ for example.
2. Find the more highly smoothed sequence $Y_8(i)$ and use it as a reference.
3. At each point *(i)*, find the square of the difference between sequences $Y_3(i)$ and $Y_8(i)$.
4. Divide the result of step 3 by the square of the $Y_8(i)$ values.
5. Get the sum of the squares and divide the sum by the number of values in the rectangular region.
6. Find the positive square root of step 5 that produces an RMS noise voltage.
7. The result is a single *estimate* of the RMS signal in the partially smoothed sequence. Each repetition of steps 1 to 6 is an additional estimate that can be averaged with the others.

The regions outside the rectangle are not included because the noise signal in this example is gated on only within the rectangle; however, the smoothing of rectangle-plus-noise spreads the rectangle itself almost as in Fig. 4-1a. All of this can be combined into a single equation for the noise ratio (NR) within the rectangle region in Fig 4-1b:

$$\mathrm{NR_{dB}} = 10\log\left[\frac{1}{49}\sum_{i=8}^{56}\left(\frac{Y_3(i) - Y_8(i)}{Y_8(i)}\right)^2\right] \quad \mathrm{dB} \quad (4\text{-}1)$$

The result is the ratio, in dB, between a three-sweep and an eight-sweep record.

Another sequence that was evaluated has seven values: $\Sigma[0.015625 + 0.09375 + 0.23438 + 0.31250 + 0.23438 + 0.09375 + 0.015625] = 1.0$. This sequence was found to give about the same results for five passes as the three-point sequence with eight passes. The spectrum plot in Fig. 4-2a compares the two methods. Mathcad finds the three-point method to be

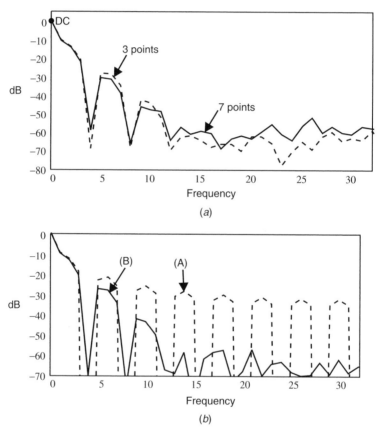

Figure 4-2 (a) Comparing a three-point sequence with a seven-point sequence in spectrum decibels. (b) Comparing the spectrum of the rectangular pulse of Fig. 4-1 before (A) and after (B) eight 3-point smoothings. (c) (1) Time domain of triangular waveform; (2) spectrum before (gray line) and after (black line) three-point smoothing of the spectrum.

consistently better at the higher frequencies. Also, the smoothing of the sequences can be in the time domain (Fig. 4-1) or the frequency domain (Fig. 4-2). For further study of smoothing methods, see, for example, [Oppenheim and Schafer 1975] and [Jenkins and Watts 1968].

We see a potential problem in Fig. 4-1a. A smoothed record that exists from 0 to $N - 1$ calls for a value of (i) in the first step that is less than zero (in a negative-time region). The final step calls for an (i) value that is greater than $N - 1$ (in a positive-time region). An excellent way to handle this is to use *circular smoothing*. When $i = -1$ is called for, use $i = N - 1$, and when $i = N$ is called for, use $i = 0$. This keeps us within

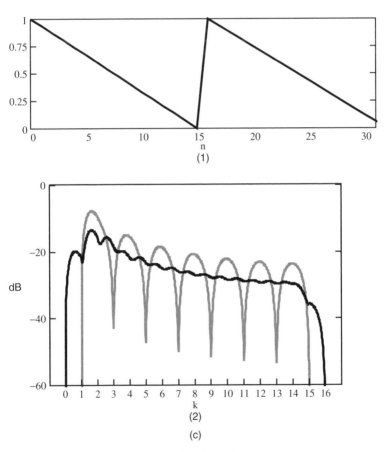

(c)

Figure 4-2 (*continued*)

the 0 to $N - 1$ region. Some additional programming steps can accomplish this, but for this example, because of the two almost zero-valued end regions, we assign the value 0.0, which is very nearly correct. Mathcad assigns 0 to unused locations, for example at $i = -1$ and $i = N$. If necessary, the two guardbands can be lengthened a little.

We see also that the sequence can be a modified time sequence, in which case the smoothing is filtering certain regions of the time domain, or it can be a modified frequency sequence, in which case certain frequency ranges can be modified. For example, the sharp edges of a band filter are softened and rounded to obtain benefits such as improved group delay near the band edges. This method is also known as *transition sampling* [Oppenheim and Schafer, 1975, pp. 250–254]. In this case the IDFT is the time domain of the modified frequency response that can be used to improve an analog filter or digital filter.

Analog mechanical and crystal filters often use the *transitional design*, which is the Bessel response to the -6-dB level for improved phase linearity, and Chebyshev beyond that, for a good shape factor. All-pass networks further improve group delay variations, at the cost of additional time delay. When networks with too much time delay are used inside an automatic gain control (AGC) loop, transient response and stability become much more difficult. We almost always avoid putting these types of filters within a fast-responding gain-control feedback loop.

The smoothing should focus on undesired rapid changes without degrading excessively the desired slower-changing signal. A good way to implement this is to use many closely spaced samples and use the three-point method no more times than necessary for adequate results. A further consideration is the rounding off at the corners, where smoothing has reduced the 3-dB width of the time or frequency response. This can often be compensated by modifying the time and frequency *scaling factors* described in Chapter 1 in a way that retains the low-amplitude guardbands at the edges of the time or frequency range. Finally, Fig. 4-2b shows the spectrum (the DFT) of the pulse of Fig. 4-1b before and after eight three-point smoothing operations. The lowpass filtering that the smoothing performs is apparent.

The smoothing algorithm is therefore a valuable addition to the toolbox. We have emphasized that if the signal pattern is too close to the edges it can alias into adjacent time or frequency regions. To prevent this and retain a *causal* (n and k from 0 to $N - 1$) result, we must delay and terminate the sequence and use smoothing and windowing sufficiently that aliasing overflows are small. A positive-only time or positive-only frequency display would show the aliasing reflecting off of the zero or maximum boundary, just as in the aliasing considered in Chapter 3. Other types of very useful windows will be studied in this chapter.

As a final example, Fig. 4-2c shows a triangular time-domain signal. Its spectrum is also shown, to which a window with 100-dB attenuation from 0 to 1 and from 15 to 16 has been added. A single 3-point smoothing is then performed on the spectrum. The spreading created by the smoothing is visible from 0 to 1 and from 15 to 16. The end zones of the smoothed response have large drop-offs at 0 and 16. Aliasing below 0 and above 16 is extremely small. This type of operation can be useful where response variations need to be smoothed out by operating directly on the spectrum. Scaling methods can establish the correct time and frequency parameters. The modified time-domain response can be found from the IDFT applied to the smoothed spectrum.

SMOOTHING REFERENCES

Jenkins, G.M., and D.G. Watts, 1968, *Spectral Analysis and Its Applications*, Holden-Day, San Francisco, CA.

Oppenheim, A.V., and R.W. Schafer, 1975, *Digital Signal Processing*, McGraw-Hill, New York.

WINDOWING

A window is a function that *multiplies* a time or frequency sequence, thereby modifying certain properties of the sequence. An example was shown in Fig. 4-1, where a rectangular two-sided time window with unity amplitude provided zero-amplitude guardbands at beginning and end segments of the sequence. The general form for a window operation $w(n)$ on a time sequence $x(n)$ of length N is

$$y(n) = w(n)x(n) \quad 0 \leq n \leq N - 1 \tag{4-2}$$

Note that the sequences $y(n)$, $w(n)$, and $x(n)$ are positive-time in the first half and negative-time in the second half, in agreement with our previously established protocol.

There are a large number of window functions that accomplish various goals. We will look at three time-domain windows that are widely used and quite useful: the rectangular window, the Hamming window, and the Hanning (also called the Hann) window. These windows, their

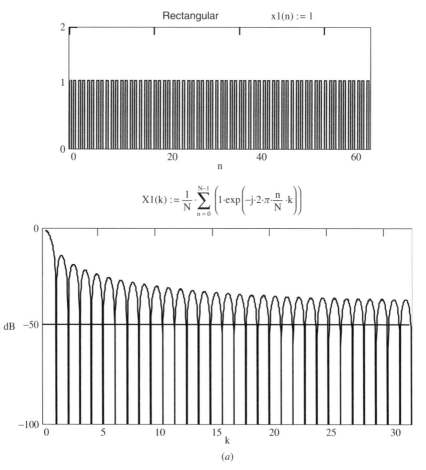

$$X1(k) := \frac{1}{N} \cdot \sum_{n=0}^{N-1} \left(1 \cdot \exp\left(-j \cdot 2 \cdot \pi \cdot \frac{n}{N} \cdot k \right) \right)$$

(a)

Figure 4-3 Three types of window: (a) rectangular; (b) Hanning; (c) Hamming.

spectral magnitudes in dB and their equations for a sequence of N positions are shown in Fig. 4-3. The spectrum magnitudes are shown for the positive-frequency halves. The negative-frequency *magnitudes* are mirror images of the positive-frequency magnitudes.

Comparing the spectra of the three, we see that the rectangular window has the sharpest selectivity at zero frequency, with a very deep notch at $k = 1$. This window is ideal when time $x(n)$ and frequency $X(k)$ values can be kept very close to exact integer values of (n) and (k), as we

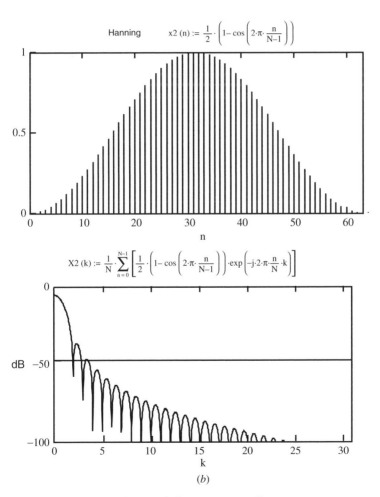

$$\text{Hanning} \qquad x2\,(n) := \frac{1}{2} \cdot \left(1 - \cos\left(2 \cdot \pi \cdot \frac{n}{N-1}\right)\right)$$

$$X2\,(k) := \frac{1}{N} \cdot \sum_{n=0}^{N-1}\left[\frac{1}{2} \cdot \left(1 - \cos\left(2 \cdot \pi \cdot \frac{n}{N-1}\right)\right) \cdot \exp\left(-j \cdot 2 \cdot \pi \cdot \frac{n}{N} \cdot k\right)\right]$$

(b)

Figure 4-3   (*continued*)

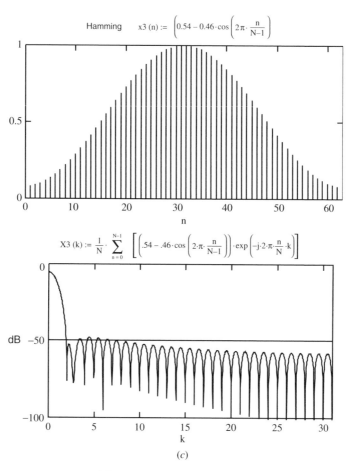

Figure 4-3   (*continued*)

discussed in connection with Fig. 3-1. In mathematical constructions we can often manage this, using scaling procedures. But if not, we see that this window has side lobe peaks that attenuate very slowly. In many practical situations, especially in digital processing of actual signal sequences with imprecise alignment or with noise contamination, this slow attenuation cannot be tolerated.

We now look at the other two windows and see that the selectivity is *widened* in the $k = 0$ region, and the first notch is at $k = 2$. This widening is a fundamental property of all non-rectangular windows and is the "cost" for the improved attenuation of the side lobes. For the Hann the value of the peak response is -38 dB at $k \approx 2.4$, for the Hammimg it is -53 dB at $k \approx 2.2$, and for the rectangular it is only -18 dB at $k \approx 2.5$. Comparing the main lobe widths in the vicinity of $X(k) = 1$, the Hann is 1.46 at -20 dB and the Hamming is 1.36 at -20 dB, which can perhaps be a worthwhile improvement.

Comparing the Hamming and the Hann, the Hamming provides deeper attenuation of the first side-lobe (which is one of its main goals) and limits in the neighborhood of 45 to 60 dB at the higher-frequency lobe peaks (another goal). For many applications this is quite satisfactory. On the other hand, the Hann is not quite as good up close but is much better at higher frequencies, and this is often preferred. In many introductory references, these two windows seem to meet the majority of practical requirements for non-integer frequency (k) values.

In the equations for the Hamming and Hann windows we see the sum of a constant term and a cosine term. There are other window types, such as the Kaiser window and its variations, that have additional cosine terms. These may be found in the references at the end of the chapter and are not pursued further in this book. These other window types are useful in certain applications, as discussed in the references.

We noted that the time-domain window sequence *multiplies* a time-domain signal sequence. In the frequency domain the spectrum of the window *convolves* with the spectrum of the signal. These interesting subjects will be explored in Chapter 5.

Figure 4-4 is a modification of Fig. 4-3 that illustrates the use of convolution in the frequency domain. Equation (4-3) contains formulas for the spectra of the windows, including frequency translation to 38.0.

$$\text{Rectangle:} \quad X_1(k) = \frac{1}{N} \sum_{n=0}^{N-1} \left[(1) \left(\exp\left(j2\pi \frac{n}{N}(38.0 - k) \right) \right) \right]$$

$$\text{Hamming:} \quad X_2(k) = \frac{1}{N} \sum_{n=0}^{N-1} \left\{ \left(0.54 - 0.56 \left[\cos\left(2\pi \frac{n}{N-1} \right) \right] \right) \right.$$

$$\left. \times \left(\exp\left(j2\pi \frac{n}{N}(38.0 - k) \right) \right) \right\} \qquad (4\text{-}3)$$

Hann: $$X_3(k) = \frac{1}{N} \sum_{n=0}^{N-1} \left\{ \frac{1}{2} \left(1 - \left[\cos \left(2\pi \frac{n}{N-1} \right) \right] \right) \right.$$

$$\left. \times \left(\exp \left(j2\pi \frac{n}{N}(38.0 - k) \right) \right) \right\}$$

The two-sided baseband signal is translated up to a center frequency of $+38.0$, where it shows up as a lower sideband and an upper sideband. The way that the Hamming dominates from 38.0 to 41.0 and the Hanning takes over starting at 42 is quite noticeable. The performance of the Hamming

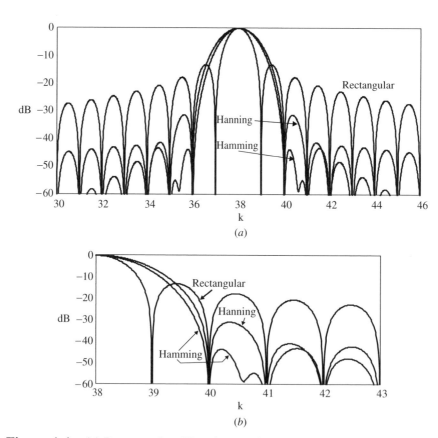

Figure 4-4 (a) Rectangular, Hanning, and Hamming windows translated to $k = 38.0$. (b) Close-up of part (a) showing window behavior between integer k values.

from 40 to 41 is also interesting. Chapter 8 shows how a single-sideband spectrum (USSB or LSSB) for these windows can be generated.

Referring again to Fig. 4-3, it is apparent that if we can stay away from $k < 2$, the Hann and Hamming are more tolerant of frequency departures from integer values. This should be considered when designing an experiment or when processing experimental data or a communication signal. If the number of n and k values can be doubled, the resolution can be improved so that after adjusting the frequency scaling factor, $k = 2$ represents a smaller actual frequency difference. If a certain positive frequency range 0 to 10 kHz is needed and an N value of 256, or 128 positive frequencies, is chosen, the resolution is 78 Hz per bin, and $k = 2$ corresponds to 156 Hz, which may not be good enough. For $N = 1024$ (512 positive), the resolution is 19.5 Hz and $k = 2$ corresponds to 39 Hz, which is a lot better.

Increasing N would also seem to make the close alignment with integer values more desirable. But in the Hamming example the sidelobe *peaks* are better than 43 dB below the $k = 0$ level, which is often good enough, and means that alignment with integer (k) values may be completely unnecessary (compare this with the rectangular window). This reduction of lobe peaks and the reduced need for integer (k) values is the major goal of window "carpentry." Note also that the $k = 0$ value is about 5 dB below the 0-dB reference level, and a gain factor of 5 dB can be included in the design to compensate.

The operations just concluded can be extended to multiple input signals. Equation (4-2) can be restated as follows:

$$
\begin{aligned}
y(n) &= w(n)(x_1(n) + x_2(n) + \cdots) \\
&= w(n)x_1(n) + w(n)x_2(n) + \cdots \\
Y(k) &= W(k) * (X_1(k) + X_2(k) \cdots) \\
&= W(k) * X_1(k) + W(k) * X_2(k) \cdots
\end{aligned}
\tag{4-4}
$$

where $*$ is the convolution operator. This means that multiplication of a window time sequence and the sum of several signal time sequences is a distributive operation, and the convolution of their spectra is also a distributive operation. Any window function performs the same operation

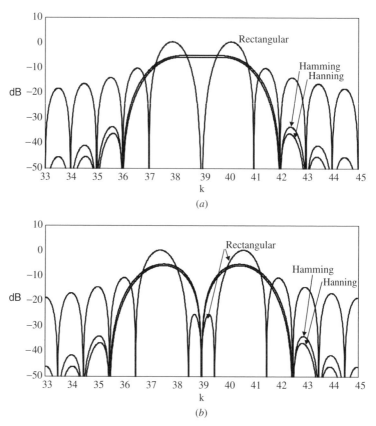

Figure 4-5 Two-tone input signal: (a) with 2 units of frequency separation; (b) with 3 units of frequency separation.

on *each* of the signal functions in the time domain and also in the frequency domain. This is mentioned because it may not be immediately obvious.

An illustration of this is shown in Fig. 4-5a for two signals at 38.0 and 40.0 (poor separation) and in Fig. 4-5b for 37.5 and 40.5 (better separation), for each of the three window functions. Using the frequency scaling factors as described previously, the resolution can be adjusted as required. At certain other non-integer close separations it will be noticed that adjacent lobe peaks interact and deform each other slightly (the reader is encouraged to try this).

The frequency conversions in Figs. 4-4 and 4-5 are exactly identical to the idealized "mixer circuit" found in radio textbooks, where a positive-frequency baseband signal and a positive-frequency local oscillator (L.O.) are multiplied together to produce upper and lower sidebands about the L.O. frequency with a suppressed L.O. frequency content. The frequency conversion itself is a second-order nonlinear process, as Eq. (4-3) confirms, but the two mixer sideband output amplitudes are each linearly related to the baseband input amplitude if the L.O. level is assumed to be constant, hence the colloquial term "linear mixer." Actual mixer circuits are not exactly linear in this manner. We also mentioned previously the two-sided baseband spectrum being translated to produce a double-sideband output. Both concepts do the same thing in the same way.

There are situations where the signal data extend over a long time period. The analysis can be performed over a set of smaller windowed time periods that intersect *coherently* so that the overall analysis is correct. The article by Harris [1978] is especially excellent for this topic and for the subject of windows in general; see also [Oppenheim and Schafer, 1975].

A frequent problem involves sudden transitions in the amplitude values between the end (or beginning) of one time sequence and the beginning (or end) of the next. This causes a degradation of the spectrum due to the introduction of excessive undesired components and also significant aliasing problems. The methods described in the smoothing section of the chapter can take care of this problem using the following guidelines: (1) create a nearly-zero amplitude guardband at each end of the sequence; (2) perform one or more three-point smoothing operations on the time and/or frequency data; (3) use scaling techniques to get the required time and frequency coverage and resolution; and (4) use a window of the type discussed in this segment to reduce the need for exact integer values of time and frequency.

We see also that the Hamming and Hanning time-domain windows are zero or almost zero at the edges, which improves protection against aliasing in the time domain. Frequency-domain aliasing is also improved, especially with the Hanning window, as the spectrum plots show. Other window types, such as the Kaiser, can be compared for these properties.

WINDOWING REFERENCES

Harris, F. J., 1978, On the use of windows for harmonic analysis with the Fourier transform, *Proc. IEEE*, Jan.

Oppenheim, A. V., and R. W., Schafer, 1975, *Digital Signal Processing*, McGraw-Hill, New York.

5

Multiplication and Convolution

Multiplication and convolution are very important operations in discrete sequence operations in the time domain and the frequency domain. We will find that there is an interesting and elegant relationship between multiplication and convolution that is useful in problem solving.

MULTIPLICATION

For the kinds of discrete time $x(n)$ or frequencies $X(k)$ of interest in this book, there are two types of multiplication. The (n) and (k) values are integers from 0 to $N - 1$. The $X(k)$ values to be multiplied are phasors that have amplitude, frequency, and phase attributes, and the $x(n)$ values have amplitude and time attributes. The Mathcad program sorts it all out. Each sequence is assumed by the software to be one realization of an infinite, steady-state repetition, with all of the significant information available in a single two-sided (n) or (k) sequence, as explained previously and mentioned here again for emphasis.

Discrete-Signal Analysis and Design, By William E. Sabin
Copyright © 2008 John Wiley & Sons, Inc.

Sequence Multiplication

One type of multiplication is the distributed sequence multiplication seen in Eq. (4-2) and repeated here:

$$z(n) = x(n)\, y(n), \quad 0 \le n < N - 1 \tag{5-1}$$

Each element of $z(n)$ is the product of each element of $x(n)$ and the corresponding element of $y(n)$. Frequently, $x(n)$ is a "weighting factor" for the $y(n)$ value. For example, $x(n)$ can be a window function that modifies a signal waveform $y(n)$. Chapter 4 showed some examples that will not be repeated here. The values $x(n)$ and $y(n)$ may in turn be functions of one or more *parameters* of (n) at each value of (n), which is "grunt work" for the computer. We are often interested in the sum $z(n)$ over the range 0 to $N - 1$ as the sum of the product of each $x(n)$ and each $y(n)$. Also, the time average or mean-square value of the sum or other statistics is important. And of course, we are especially interested in $Z(k)$, the spectrum of $z(n)$, $Y(k)$, the spectrum of $y(n)$, and $X(k)$, the spectrum of $x(n)$.

Figure 5-1 is another example of frequency conversion by using this kind of multiplication. A time sequence $x(n)$ at a frequency $k = 4$ and a time sequence $y(n)$ at frequency $k = 24$ are multiplied term-by-term to get the time sequence for the product.

The DFT then finds the two-sided phasor spectrum. The one-sided spectrum is found by adding the phasors at 108 and 20 to get the positive cosine at 20, and adding the 100 and 28 phasors to get the negative cosine term at 28. See Fig. 2-2a to confirm these results, and note the agreement with the equation for $z(i)$ in Fig. 5-1. As we said before, the input frequencies 24 and 4 disappear, but if one input amplitude is held constant, the product is linear with respect to variations in the other input amplitude.

Polynomial Multiplication

The other kind of sequence multiplication is polynomial multiplication, which uses the distributive and associative properties of algebra. An example is shown in Eq. (5-2).

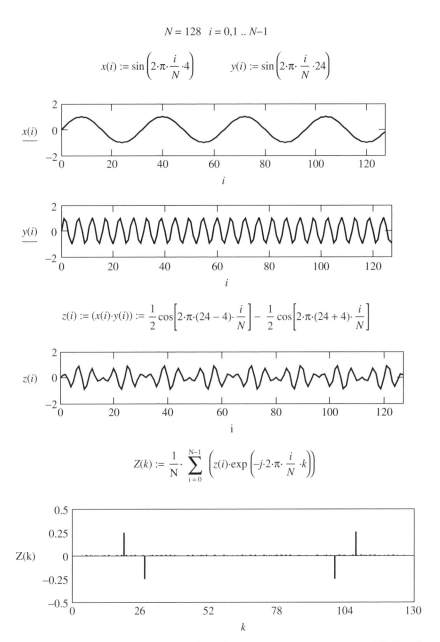

Figure 5-1 Frequency conversion through term-by-term multiplication of two time sequences.

$$z(x,y) = (x_1 + x_2 + \cdots + x_\alpha)(y_1 + y_2 + y_3 + \cdots + y_\beta)$$

$$= \sum_y \left(\sum_x z(x) \right) z(y)$$

$$z(i) = x(i)(y_1 + y_2 + y_3 + \cdots + y_\beta) \tag{5-2}$$

$$= x(i) \sum_j y(j)$$

Each term of the x sequence is multiplied by the sum of the terms in the (y) sequence to produce each term in the (z) sequence, which then has α terms. Or, each term of the (y) sequence is multiplied by the sum of the terms in the (x) sequence to get β terms. Or, each term in the first sequence is multiplied by each term in the second sequence and the partial products are added. In all of these ways, the (z) sequence is the polynomial product of the (x) and (y) sequences. The sum of $z(i)$ is the "energy" in (z). This, divided by the "time" duration, is the "average power" in $z(i)$. If the sum of (x) or the sum of (y) equals zero, the product is zero, as Eq. (5-2) clearly indicates. For certain parts of the range of $x(i)$ and $y(j)$ the product can usually be nonzero. In the familiar arithmetic multiplication, the (x) and (y) terms are "weighted" in descending powers of 10 to get the correct answer: for example,

$$\begin{aligned}
8734 \cdot 4356 &= (8000 + 700 + 30 + 4) \cdot (4000 + 300 + 50 + 6) \\
&= 8000 \cdot 4000 + 8000 \cdot 300 + 8000 \cdot 50 + 8000 \cdot 6 \\
&\quad + 700 \cdot 4000 + 700 \cdot 300 + 700 \cdot 50 + 700 \cdot 6 \\
&\quad + 30 \cdot 4000 + 30 \cdot 300 + 30 \cdot 50 + 30 \cdot 6 \\
&\quad + 4 \cdot 4000 + 4 \cdot 300 + 4 \cdot 50 + 4 \cdot 6 \\
&= 38,045,304
\end{aligned} \tag{5-3}$$

Polynomial multiplication $A(x) B(y)$ is widely used, including in topics in this and later chapters. It shows how each item in sequence $A(x)$ affects a *set* of items in sequence $B(y)$. It is equivalent to a double integration or double summation such as we might use to calculate the area of a two-dimensional figure. Figure 5-2 is a simple example. More complicated geometries require that the operation be performed in segments and the partial results combined.

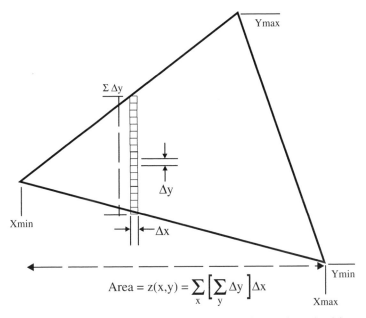

Figure 5-2 Example of polynomial multiplication using double summation to find the area of a figure.

CONVOLUTION

Convolution is a valuable tool for the analysis and design of communications systems and in many other engineering and scientific activities. Equation (5-4) is the basic equation for discrete-time convolution.

$$y(n) = x(m) * h(m) = \sum_{m=-\infty}^{+\infty} [x(m) \, h(n-m)] \qquad (5\text{-}4)$$

where $*$ is the convolution operator and $y(n)$, $x(m)$, and $h(m)$ can all be the complex-valued discrete-time *sequences* I and jQ that we considered carefully in Chapter 1. Note that $x(m)$, $h(m)$, and $y(n)$ are in the time domain, but they can also be complex $Y(k)$, $X(k)$, and $H(k)$ in the frequency domain with magnitude and phase attributes. Also, all three can have different amplitude scale factors, on the same graph or on separate graphs. We focus initially on the time domain.

Equation (5-4) appears to be simple enough, but actually needs some careful study and practice to develop insight and to assure correct answers.

Sequence $x(m)$ is a "signal" input time-domain sequence that extends in "time" from $m = -\infty$ to $m = +\infty$. In practical problems this sequence is assumed to have "useful" amplitude only between two specific limits, m(min) and m(max). Sequence $h(m)$ refers to a "system function," also a time-domain sequence that is assumed to have useful limits from m(min) to m(max), which may not be the same limits as the limits for $x(m)$. Sequence $h(n - m)$ is $h(m)$ that has had two operations imposed: 1: $h(m)$ has been "flipped" in time and becomes $h(-m)$, and 2: $h(-m)$ has been shifted to the right (n) places, starting from an initial value of (n) determined by the nature of the problem, whose value is now $h(n - m)$. The expression "fold and slide" is widely used to describe this two-part operation.

One reason for the fold and slide of $h(m)$ to $h(n - m)$ is that we want the leading edge in time of $h(m)$ (the response) to encounter the leading edge in time of $x(m)$ (the excitation) as we start the sliding operation (increasing n) from starting (n) to final (n). This retains the time coordination between the two sequences. Another reason is that this procedure leads to a valuable concept that we will demonstrate later in this chapter.

The term $h(n - m)$ is often referred to as an *impulse response*. Each impulse $x(m)$ applied to a network is processed by some impulse response function $h(n - m)$ to produce an output impulse $y(n)$, which is the "value" of the convolution for that (n). Note the summation sign in Eq. (5-4).

Refer to Fig. 5-3. In part (a), $h(m)$ and $x(m)$ are at first located side by side. This is a convenient starting point. In part (b), $h(m)$ is flipped and pivoted about the $m = 0$ point and becomes $h(-m)$. At each positive increment of (n), $h(n - m)$ is advanced one position to the right. We now calculate the product of $x(m)$ and $h(n - m)$ at each value of (m), and add these products from x(min) to x(max) to get $y(n)$, the convolution value of $x(m)$ and $h(m)$ at that value of (n). We continue this shift–multiply–add procedure until the value of the sum of products (the convolution value) becomes negligible.

Make sure that the data space (0 to $N - 1$) is enough that the total convolution sum can be completed with no significant loss of data. If the $x(m)$ sequence has Lx values and the side-by-side $h(m)$ sequence has Lh values, the $y(n)$ sequence should have at least $Lx + Lh + 1$ values prior to the beginning of overlap. The convolution operation creates a smoothing and stretching operation on the data $y(n)$ which is not obvious in Figs. 5-3 and 5-4 but is more visible if we look ahead to Fig. 5-7.

The convolution sequence that has been generated has properties that will be discussed subsequently. This procedure should be compared to the

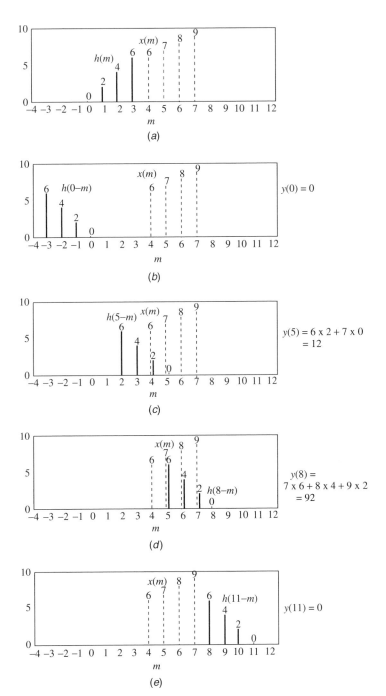

Figure 5-3   Various stages of discrete convolution.

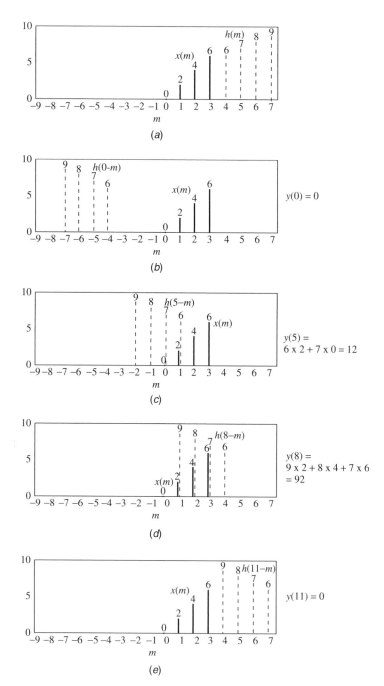

Figure 5-4   Altered stages of discrete convolution.

convolution of continuous functions, in which certain discontinuities in the functions and their overlaps require integrations over adjacent areas which are then tacked together end-to-end to get the final result. Our example sequences will not have this problem because of their discrete nature, although this may not always be true.

This process is an example of the polynomial multiplication shown in Eq. (5-2). That is, at a particular value of (n), the product of each $x(m)$ and the corresponding $h(n-m)$ is formed. These partial products are summed over the range $m(\min)$ to $m(\max)$ to get the convolution value $y(n)$. This process is repeated for each value of (n).

To illustrate this discussion more numerically, we see again in Fig. 5-3 the convolution procedure for the two discrete *sequences* shown in part (a). The $m=0$ location is chosen arbitrarily as shown, but in principle it can be anywhere between $-\infty$ and $+\infty$. In agreement with our previous assumptions, the process shown is significant between $0 \le n \le N-1$, where N is chosen to suit the problem. A convenient approach is to start the $h(m)$ sequence at a point labeled "0".

In Fig. 5-3b the $h(m)$ sequence is reversed (or folded) and pivoted around the zero location, shown at the left end of $h(m)$, and becomes $h(-m)$. That is, each m (positive location) becomes m (negative location). After pivoting, $h(-m)$ can be moved three places to the right, where it becomes $h(3-m)$, just one "bin" away from $x(m)$.

In part (c) the $h(3-m)$ sequence slides to the right two places ($n=5$) and overlaps the first two positions of the $x(m)$ sequence. The value of the convolution for $n=5$ is $6{\cdot}2+7{\cdot}0=12$.

In part (d) the value of (n) is 8 and the convolution sum is $7{\cdot}6+8{\cdot}4+9{\cdot}2=92$. At $n=11$ the overlap of $h(n-m)$ and $x(m)$ is zero and the convolution sum for $n=11$ is zero.

In Fig. 5-4 $x(m)$ and $h(m)$ are interchanged; $x(m)$ is response and $h(m)$ is the signal, with no differences in the convolution values, which is an interesting facet of convolution that is the "commutative" property. We see also that convolution involves two time-domain sequences or two frequency-domain sequences and the convolution value (sum of products) has the dimension of energy.

Discrete Circular Convolution

In the sequences used in convolution, any empty integer locations are filled with zeros by Mathcad. The combined length L of the two sequences $x(m)$

and $h(m)$ must provide an initial separation of at least one (m). If we need to reduce L in order to reduce the sequence length, the method of circular convolution can be used, as illustrated in Fig. 5-5. Note that the amplitude scales and the number of samples for $x(m)$ and $h(m)$ can be different. The following steps are executed.

(a) $x(m)$ is plotted.

(b) $h(m)$ is plotted.

(c) $h(m)$ is flipped about the $m=0$ point on the m-axis. The $h(m)$ line at $m=+2$ (length equals zero) now appears at $m=-2$, which is the same as $m=-2+16=+14$. The line that was at $m=+11$ (length 9) now appears at $m=-11$, which is the same as $m=-11+16=+5$.

(d) We do not start the sum of products (convolution) of overlapping $x(m)$ and $h(-m)$ at this time.

(e) The $h(-m)$ sequence advances 2 places to $h(2-m)$. The line at $m=14$ (length zero) moves to $m=16$, which is identical to $m=0$. The $h(n-m)$ and $x(m)$ sequences are now starting to intersect at $m=0$.

(f) The process of multiply, add, and shift begins. The result for $(11-m)$ is shown for which the convolution value is 224. For values greater than $(18-m)$ the convolution value is 0 because x and h no longer intersect.

As we can see, this procedure is confusing when done manually (we do not automate it in this book), and it is suggested that the side-by-side method of Figs. 5-3 and 5-4 be used instead. Adjust the length $L=[x(m)+h(m)+1]$ as needed by increasing $N(=2^M)$, and let the computer perform Eq. (5-4), the simple sum of products of the overlapping $x(m)$ and $h(m)$ amplitudes. The lines of zero amplitude are taken care of automatically. Design the problem to simplify the process within a reasonable length L.

Time and Phase Shift

In Fig. 5-6 a sequence $x(m)$ is shown. A single $h(3)$ is also shown and $h(-3)$ is the flip from $+3$ to -3. The amplitude of $h(m)$ is 2.0. If we now suddenly move $h(-3)$ forward six places to become $h(6-3)$ for an $n=6$, we get $y(3)$, which is a copy of $x(6-3)$, shifted to the right three

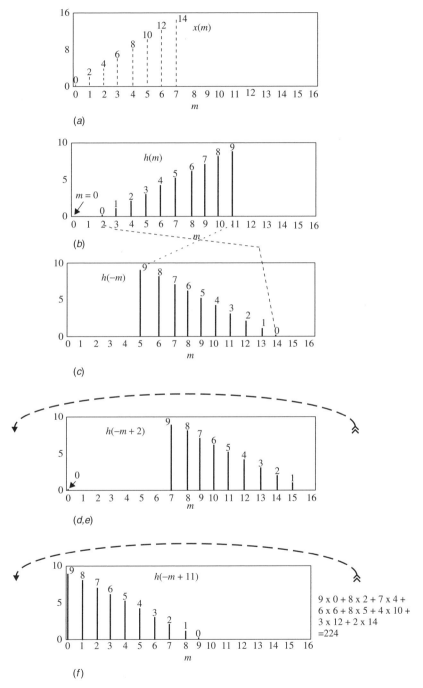

Figure 5-5 Circular discrete convolution.

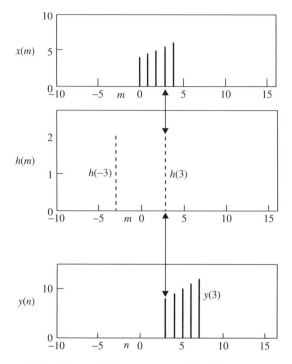

Figure 5-6 Time-shift property of a discrete sequence.

places and amplified by a gain factor of 2. The $x(m)$ sequence has been suddenly moved forward (*advanced*) in time by three units, to become $y(3) = x(m+3)$. If $x(m)$ is $360°$ of a periodic sequence, its phase has been advanced (forwarded) by $360 \cdot 3/5 = 216°$. If $y(n)$ is the un-shifted sequence, the shifted sequence $y(n+u)$ is

$$y(n+u) = y(n)\, e^{j2\pi(u/N)}$$
$$= y(n)\left[\cos\left(2\pi \tfrac{u}{N}\right) + j\sin\left(2\pi \tfrac{u}{N}\right)\right] \quad (5\text{-}5)$$

where the positive sign of the exponential indicates a positive (counterclockwise) phase rotation and also a positive time advance as defined in Chapter 1 and Fig. 1-5. In other words, $y(n)$ has suddenly been advanced forward three time units.

We can more easily go directly from $x(m)$ to $y(n)$ by just advancing $x(m)$ three places to the right, which according to Eq. (5-5) is the time-advance principle (or property) applied to a discrete sequence.

DFT and IDFT of Discrete Convolution

If we take the discrete convolution of Eq. (5-4) and apply the DFT of Eq. (1-2), we can get two useful results. The first is that the discrete *convolution* $y(m) = x(m) * h(m)$ in the time-domain transforms via the DFT to the discrete *multiplication* $Y(k) = X(k)H(k)$ in the frequency domain in the manner of the sequence multiplication of Eq. (5-1). That is, $Y(k)$ is the "output" of $X(k)H(k)$ at each k. If $X(k)$ is the voltage or current spectrum of an input signal and $H(k)$ is the voltage or current frequency response of a *linear* filter or *linear* amplifier, for example, then $Y(k)$ is the voltage or current spectrum of the output of the linear system.

The second result is that the IDFT [see Eq. (1-8)] of the discrete convolution $Y(k) = X(k) * H(k)$ produces the discrete product $y(n) = x(n)h(n)$. This type of convolution is often used, and one common example is frequency translation, where a baseband signal with a certain spectrum is convolved with an RF signal that has a certain RF spectrum to produce an output RF spectrum $Y(k)$. The IDFT of $Y(k)$ is $y(n)$, the time-domain output of the system, which is often of great interest.

It is interesting to see how the DFT of convolution in the time domain leads to the product in the frequency domain. We start with the double summation in Eq. (5-6), which is the DFT [Eq. (1-2)] of the discrete convolution [Eq. (5-4)].

$$Y(k) = \frac{1}{N} \sum_{n=0}^{N-1} \underbrace{\sum_{m=-\infty}^{+\infty} \left[x(m)h(n-m) \right]}_{\text{convolution}} e^{-j2\pi k(n/N)} \qquad (5\text{-}6)$$

$$\underbrace{\phantom{Y(k) = \frac{1}{N} \sum_{n=0}^{N-1} \sum_{m=-\infty}^{+\infty} \left[x(m)h(n-m) \right] e^{-j2\pi k(n/N)}}}_{\text{DFT}}$$

We will modify this, with no resulting error, so that the limits on both summations extend over an infinite region of n and m, with unused locations set equal to 0, and with n and m interchanged in the summation symbols.

$$Y(k) = \frac{1}{N} \sum_{m=-\infty}^{m=+\infty} \sum_{n=-\infty}^{n=+\infty} \left[x(m)h(n-m) \right] e^{-j2\pi k(n/N)} \qquad (5\text{-}7)$$

Since $x(m)$ is not a function of n, it can be associated with the first summation (again with no error) and the exponential term is associated only with n.

$$Y(k) = \frac{1}{N} \sum_{m=-\infty}^{m=+\infty} x(m) \left[\sum_{n=-\infty}^{n=+\infty} h(n-m)e^{-j2\pi k(n/N)} \right] \qquad (5\text{-}8)$$

We now apply the well-known time delay theorem of the Fourier transform to the second summation in brackets [Carlson, 1986, pp. 52 and 655; Schwartz, 1980, pp.72– 73]. The DFT of this expression is

$$\text{DFT of } h(n-m) = H(k)e^{-j2\pi k(m/N)} \qquad (5\text{-}9)$$

and Eq. (5-8) can be rewritten as

$$Y(k) = \frac{1}{N} \sum_{m=-\infty}^{m=+\infty} x(m) H(k) e^{-j2\pi k(m/N)}$$

$$= \frac{1}{N} \underbrace{\sum_{m=-\infty}^{m=+\infty} x(m)\, e^{-j2\pi k(m/N)}}_{=X(k)}\, H(k) \qquad (5\text{-}10)$$

$$= X(k)\, H(k)$$

The same kind of reasoning verifies that the IDFT of $Y(k)=X(k)*H(k)$ and leads to $y(n)=x(n)h(n)$. These kinds of manipulations of sums (and integrals) are often used and must be done correctly. The references at the end of this chapter accomplish this for this application.

Figure 5-7 illustrates the general ideas for this topic, which we discuss as a set of steps (a) to (h), and also confirms Eq. (5-10):

(a) This is $x(m)$, a nine point rectangular time-domain sequence.
(b) This is $h(m)$, a 32-point time-domain sequence with an exponential decay.

(c) The convolution of (a) and (b) using the basic convolution equation (5-4), $y(n) = x(m) * y(m)$. We see the familiar smoothing and stretching operation that the convolution performs on $x(m)$ and $h(m)$. Convolution needs the additional time region for correct results, as noted in the equation in part (c) and in Eq. (5-6).

(d) This step gets the convolution of $x(m)$ and $h(m)$ and also the spectrum (DFT) of the convolution in one step using the double summation of Eq. (5-6).

(e) See step (f).

(f) Steps (e) and (f). These steps get the DFT spectrum $X(k)$ of $x(m)$ and spectrum $H(k)$ of $h(m)$ using the DFT in Eq. (1-2).

(g) The product $X(k)H(k)$ is the spectrum $Z(k)$ of the "output". This product is the sequence multiplication described in Eq. (5-1). The additional factor N will be explained below. Note that the spectrum of the output in part (g) is identical to the spectrum of the output found in part (d). That is, $X(k)H(k) = $ DFT of $x(m) * h(m)$, and the IDFT of $X(k) * H(k) = x(m)h(m)$, not shown here.

(h) The IDFT in Eq. (1-8) produces the sequence $z(n)$, which is identical to the convolution $x(m) * h(m)$ that we found in part (c).

In part (g) we introduced the factor N. If we look at the equation in part (d) we see a single factor $1/N$. But the product of $X(k)H(k)$ in part (g) produces the factor $(1/N)^2$. A review of parts (e) and (f) verify this. This produces an incorrect scale factor in part (h), so we correct part (g) to fix this problem. There are different conventions used for the scale factors for the various forms of the DFT and IDFT. The correction used here takes this problem into account for the Bracewell (see Chapter 1) conventions that we are using. This action does not produce an error. it makes the "bookkeeping" correct, and the Mathcad worksheet takes the correct action as needed. These mild "discrepancies" can show up, and need not create concern.

In Fig. 5-7 the convolution in parts (c) and (h) has a sharp peak at location $n = 8$. When the signal is mildly contaminated with noise, this is a good location to place a detector circuit that creates a synchronizing pulse. Convolution is frequently employed in this manner: for example, in radar receivers [Blinchikoff and Zverev, 1971, Chap. 7].

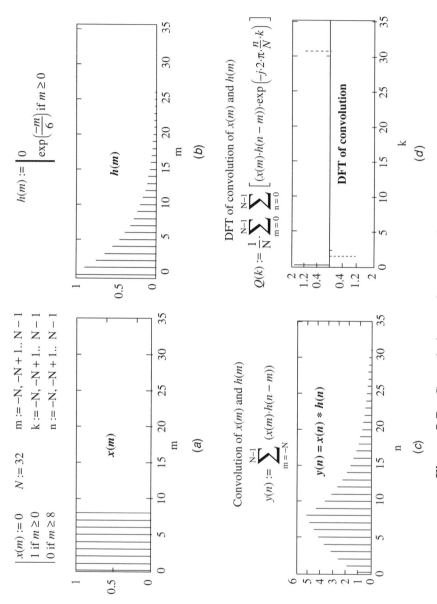

Figure 5-7 Convolution and spectra of two sequences.

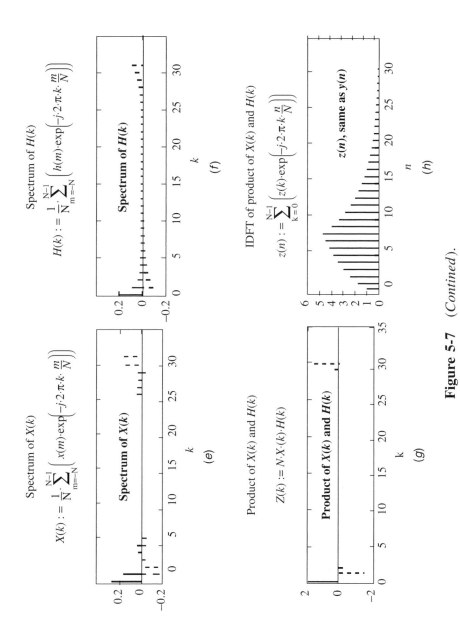

Figure 5-7 (*Continued*).

93

DECONVOLUTION

The inverse of convolution is deconvolution, where we try to separate the elements that are convolved into their constituent parts. There are some interesting uses for this, including de-reverberation, estimation of speech parameters, restoration of acoustic recordings, echo removal, video analysis, and nonlinear systems. This introductory book cannot get involved in this advanced subject, but the references, in particular [Oppenheim and Schafer, 1999, Chap. 10] and [Wikipedia] provide readable introductions. In the example of Fig. 5-7, the results in parts (c) and (h) are related to the inputs $x(m)$ and $h(m)$ in parts (a) and (b), but it may be difficult or impossible to say that these particular $x(m)$ and $h(m)$ are the only two functions that can produce the results in parts (c) and (h). When the input functions include random properties, the difficulty is compounded, and this is a major problem with deconvolution. An effort is made to find something that is not random that can be a basis for deconvolution. One example is the dynamics of an antique phonograph recording machine. Advanced statistical methods try to perform deconvolution in a noise-contaminated environment [Wikipedia].

REFERENCES

Carlson, A.B., 1986, *Communications Systems*, 3rd ed., McGraw-Hill, New York.

Schwartz, M., 1980, *Information Transmission, Modulation and Noise*, 3rd ed., McGraw-Hill, New York.

Blinchikoff, H.J., and A.I. Zverev, 1976, *Filtering in the Time and Frequency Domains* Wiley & Sons, New York.

Oppenheim, A.V., and R.W. Schafer, *Digital Signal Processing*, 1976, McGraw-Hill, New York.

Wikipedia [http://en.wikipedia.org/wiki/Deconvolution].

6

Probability and Correlation

The ideas to be described and illustrated in this chapter are introduced initially in terms of deterministic, discrete, eternal steady-state sequences, conforming to the limited goals of this introductory book. Then small amounts of additive noise (as we saw in Chapter 4) are added for further analysis. Some interesting and useful techniques will be introduced, and these ideas are used in communication systems analysis and design. This material, plus the References, will help the reader to get started on more advanced topics, but first we want to review (once more) briefly an item from Chapter 1 that we will need.

Where we perform a summation Σ from 0 to $N-1$, we assume that all of the significant signal and noise energy that we are concerned with lies within, or has been confined to, those boundaries. This also validates our assumptions about the steady-state repetition of sequences. For this reason we stipulate that all signals are "power signals" (repetitive) and not "energy signals" (nonrepetitive). In Chapter 3 we looked at aliasing and spectral leakage and ways to deal with them, and that helps to assure our reliance on 0 to $N-1$. We can increase N by 2^M ($M = 2, 3, 4, \cdots$) and

Discrete-Signal Analysis and Design, By William E. Sabin
Copyright © 2008 John Wiley & Sons, Inc.

adjust scaling factors (Chapter 1) as needed to assure "adequate" cover-age and resolution of time and frequency. The smoothing and windowing methods in Chapter 4 are helpful in reducing time and frequency require-ments without deleting important data. This is where the "art" of approx-imation and "practical" engineering get involved, and we delay the more advanced mathematical considerations for consideration "down the road."

Also, small quantities of noise can very quickly take us into the com-plexities of statistical analysis. We are for the most part going to avoid this arena because it is does not suit the limited purposes of this book. For this reason we are going to be somewhat less than rigorous in some of the embedded-noise topics to be covered. However, we will have brief contacts that will give us a "feel" for these topics.

PROPERTIES OF A DISCRETE SEQUENCE

Expected Value of $x(n)$

In Fig. 6-1a, a discrete time sequence (256 points) is shown from $n = 0$ to $N-1$. The first (positive) half is from 0 to $N/2-1$, and the second (negative) half is from $N/2$ to $N-1$, as we saw in Figs. 1-1d and 1-2b. For an infinite $x(n)$ sequence, the quantity $E[x(n)]$, the statistical *expected* value, also known as the *first moment* [Meyer, 1970, Chap. 7], is

$$E[x(n)] = \sum_{n=-\infty}^{\infty} x(n)p(x(n)) \tag{6-1}$$

where $p(x(n))$ is the *probability*, from 0.0 to $+1.0$, of occurrence of a particular $x(n)$ of an infinite sequence. In deterministic sequences from 0 to $N-1$ such as Fig. 6-1a each $x(n)$ is assumed to have an equal probability $1/N$ from 0 to $N-1$ of occurrence, the amplitudes of $x(n)$ can all be different, and Eq. (6-1) reverts to the time-average value $\langle x(n) \rangle$ from 0 to $N-1$, which we have been calling the time-domain dc value, and which is also identical to the zero frequency $X(0)$ value for the $X(k)$ frequency-domain sequence. Note that the angular brackets in $\langle x(n) \rangle$ refer to a *time average*. We have reasonably assumed that expected value and

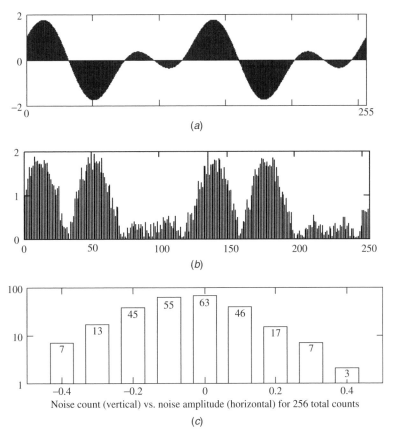

Figure 6-1 Signal without noise (a), with noise and envelope detected (b), and a sample histogram of the envelope-detected noise alone (c).

time-average value are the same for a repetitive *deterministic* sequence with little or no aliasing:

$$X(0) = E[x(n)] = \langle x(n) \rangle = \frac{1}{N} \sum_{n=0}^{N-1} x(n) \qquad (6\text{-}2)$$

all of which are zero in Fig. 6-1a.

Figure 6-1b introduces a small amount of *additive noise* voltage $\varepsilon(n)$ to each $x(n)$, so called because each $x(n)$ is the same as in Fig. 6-1a but with noise $\varepsilon(n)$ that is *added* to $x(n)$, not *multiplied* by $x(n)$ as it would be in various nonlinear applications. The noise voltage itself, as found

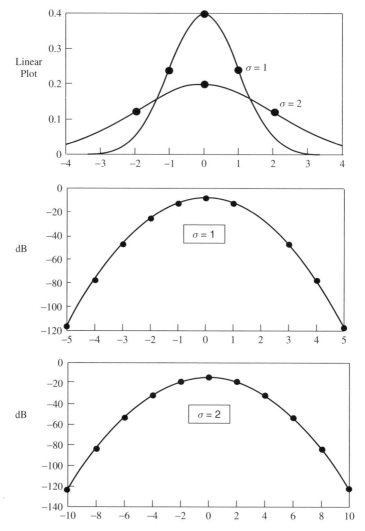

Figure 6-2   Normal distribution for $\sigma = 1$ and 2.

in communications systems, is assumed to have the familiar bell-shaped Gaussian (also called) amplitude density function shown in the *histogram* in Fig. 6-1c and in more detail in Figs. 6-2 and 6-3 [Schwartz, 1980, Chap. 5]. See also Chapter 7 of this book for a more detailed discussion of noise and its multiplication.

The noise has random amplitude. For this discussion, the signal plus noise $[x(n) + \varepsilon(n)]$ is measured by an ideal (linear) full-wave *envelope detector* that delivers the value $|x(n) + \varepsilon(n)|$ at each (n), where the vertical bars imply the positive magnitude. In Fig. 6-1b the $\varepsilon(n)$ noise voltage is a small fraction of the signal voltage. In each calculation of the sequence $x(n)$ a new sequence $\varepsilon_x(n)$ is also generated. Equation (6-3) finds the positive magnitude of $[x(n) + \varepsilon(n)]$.

$$E[|(x + \varepsilon)n|] \approx \langle |(x + \varepsilon)n| \rangle \approx \frac{1}{N} \sum_{n=0}^{N-1} [|(x + \varepsilon)n|] \qquad (6\text{-}3)$$

This is an example of the *wide-sense stationary* process, in which the expected value and the time average of a very long single record or the average of *many* medium-length records are the same and the autocorrelation value (discussed later in the chapter) depends only on the time shift τ [Carlson, 1986, Chap. 5.1].

The contribution of the sequence $\varepsilon(n)$ to the detector output is a *random variable* that fluctuates with each solution of Eq. (6-3). The *histogram* approximation in Fig. 6-1c for one of the many solutions shows a set of amplitude values of $\varepsilon(n)$, arranged in a set of discrete values on the horizontal scale, with the number of times that value occurs on the vertical scale. The total count is always 256, one noise sample for each value of n from 0 to 255. The count within each rectangle, divided by 256, is the fraction of the total occurring in that rectangle. This is a simple *normalization* procedure that we will use [see also Schwartz, 1980, Fig. A-9]. The averaging of a large number of these data records leads to an *ensemble* average. A single very long record such as $N = 2^{16} = 65,536$ using Mathcad is also a very good approximation to the expected value.

Average Power

The *average power*, also known as the *expected value* of $x(n)^2$ and also as the *time-average value* $\langle x(n)^2 \rangle$ of the deterministic (no noise) sequence in Fig. 6-1a is 1.00 W into 1.0 ohm, found in Eq. (6-4)

$$P_{\text{av}} = E[x(n)]^2 = \langle x(n)^2 \rangle = \frac{1}{N} \sum_{n=0}^{N-1} x(n)^2 = 1.00 \text{ W} \qquad (6\text{-}4)$$

This includes any power due to a dc component in $x(n)$, which in Fig. 6-1a is zero. The power in the envelope-detected signal is also 1.00 W. The total power in both cases must be identical because this ideal full-wave envelope detector is assumed to be 100% efficient. In Fig. 6-1b with added noise, the power is

$$P_{av} \approx \frac{1}{N} \sum_{n=0}^{N-1} [x(n) + \varepsilon(n)]^2$$

$$\approx \frac{1}{N} \sum_{n=0}^{N-1} \left[(x(n))^2 + 2x(n)\varepsilon(n) + (\varepsilon(n))^2 \right] \qquad (6\text{-}5)$$

$$\approx \frac{1}{N} \sum_{n=0}^{N-1} \left[(x(n))^2 + (\varepsilon(n))^2 \right] \approx 1.024 \text{ W} \qquad \text{(on average)}$$

and the additional 0.024 W is due to noise alone.

In most situations, the expected value of the product $[2x(n)\varepsilon(n)]$ is assumed to be zero because these two are, on average, statistically independent of each other. In this example with additive noise and full-wave ideal envelope detection we will be slightly more careful. Comparing noise and signal within the linear envelope detector, and assuming that the signal is usually much greater than the noise, the noise and signal are, as a simplification, in phase half of the time and in opposite phase half of the time. P(n) can then be approximated as

$$P(n) \approx [x(n) + \varepsilon(n)]^2 = x(n)^2 + 2x(n)\varepsilon(n) + \varepsilon(n)^2$$

$$\approx [x(n) - \varepsilon(n)]^2 = x(n)^2 - 2x(n)\varepsilon(n) + \varepsilon(n)^2 \qquad (6\text{-}6)$$

$$\text{average value of } P(n) \approx x(n)^2 + \varepsilon(n)^2$$

which is the same as Eq. (6-5).

Individual calculations of this product in Eq. (6-5) vary from about +20 mW to about −20 mW, and the average approaches zero over many repetitions. A single long record $N = 2^{16} = 65{,}536$ produces about ±1.0 mW, and $P_{av} \approx 1.024$ W.

The P_{av} result is then the expected value of signal power plus the expected value of noise power. This illustrates the interesting and useful fact that linear systems have *superposition* of average or expected power values that are independent (uncorrelated, see later in this chapter and Chapter 7). This P_{av} is a random variable > 0 that has an average value for a large number of repetitions or possibly for one very long sequence. Numerous repeats of Eq. (6-5) converge to values "close" to 1.024 W. In dB the ratio of desired signal power to undesired noise power is

$$\frac{S}{N} \approx 10 \log \frac{1.0}{0.024} \approx 16.2 \text{ dB} \tag{6-7}$$

We are often interested in the ratio $(S+N)/N = 1 + (S/N) \approx 16.3$ dB in this example not much different.

   This exercise illustrates the importance of averaging many calculation results when random noise or other random effects are involved. A single calculation over a single very long sequence may be too time consuming. Advanced texts consider these random effects in more excruciating detail.

Variance

Signals often have a dc component, and we want to identify separately the power in the dc component and the power in the ac component. We have looked at this in previous chapters. Variance is another way to do it in the time domain, especially when $x(n)$ includes an additive random noise term $\varepsilon(n)$, and is defined as

$$V\left(x'(n)\right) = \sigma_x^2 = E\left[x'(n) - \langle x'(n)\rangle\right]^2$$
$$= E\left(x'(n)^2\right) - \left[E\left(x'(n)\right)\right]^2 \tag{6-8}$$
$$= \langle x'(n)^2\rangle - \langle x'(n)\rangle^2$$

where $x' = x + \varepsilon$, $V(x'(n))$ is the expected or average value of the square of the entire waveform minus the square of the dc component, and the result is the average ac power in $x'(n)$. The distinction between the *average-of-the-square* and the *square-of-the-average* should be noted. The positive root $\sqrt{V(x(n))}$ is known as σ_x, the *standard deviation* of x, and has an ac rms "volts" value which we look at more closely in the next topic.

A dozen records of the noise-contaminated signal using Eq. (6-8), followed by averaging of the results, produces an ensemble average that is a more accurate estimate of the signal power and the noise power. An example of variance as derived from Fig. 6-1b, using Eq. (6-8), is shown in Eq. (6-9).

$$\text{average of the square} = E[x(n)^2] = \frac{1}{N} \sum_{n=0}^{N-1} (x(n) + \varepsilon(n))^2 = 1.0224$$

$$\text{square of the average} = (E[x(n)])^2 = \left| \frac{1}{N} \sum_{n=0}^{N-1} (x(n) + \varepsilon(n)) \right|^2 = 0.6495$$

$$\text{variance} = \text{average of the square} - 1\text{square of the average}$$

$$= 1.0224 - 0.6495 = 0.3729 \qquad (6\text{-}9)$$

$$\sigma = \sqrt{\text{variance}} = 0.6107 \text{ Vrms ac}$$

We point out also that various modes of data communication have special methods of computing the power of signal waveforms, for example, Understanding the Perils of Spectrum Analyzer Power Averaging, Steve Murray, Keithley Instruments, Inc., Cleveland, Ohio.

GAUSSIAN (NORMAL) DISTRIBUTION

This probability density function (PDF) is used in many fields of science, engineering, and statistics. We will give a brief overview that is appropriate for this introductory book on discrete-signal sequence analysis (see [Meyer, 1970, Chap. 9] and many other references). The noise contamination encountered in communication networks is very often of this type. The form of the normal curve is

$$g(m) = \frac{1}{\sqrt{2\pi}\sigma} \exp\left[-\frac{1}{2} \left\{ \frac{m - \mu}{\sigma} \right\}^2 \right] \qquad -\infty \leq m \leq +\infty \quad (6\text{-}10)$$

Note that exp(x) and e^x are the same thing. The μ term is the value of the offset of the peak of the curve from the $m = 0$ location (a positive value of μ corresponds to a shift to the right). The σ term is the standard deviation previously mentioned. Values of $g(m)$ for n outside the range of $\pm 4\sigma$ are very much smaller than the peak value and can often (but not always) be ignored.

Figure 6-2 shows two normal curves with σ values 1 and 2 and $\mu = 0$. In this figure, the discrete values of m are finely subdivided in 0.01 steps to give continuous line plots. An examination of Eq. (6-10) shows that when $m = 0$ and $\mu = 0$, the peak values of $g(m)$ are approximately 0.4 and 0.2, respectively. When m $= \pm 1$ and $\sigma = 1$, the large dots on the solid curve are located at m $= \pm 1$; similarly for $\sigma = 2$ on the dashed curve. The horizontal markings therefore correspond to integer values of σ.

Figure 6-2 also displays dB values for $\sigma = 1$ and 2, which can be useful for those values of σ. Note the changes in horizontal scale. Equation (6-10) can be easily calculated in the Mathcad program for other values of μ and σ, and the similarities and differences are noticed in Fig. 6-2.

CUMULATIVE DISTRIBUTION

The plots in Fig. 6-2 are *probability density functions* (PDFs) [Eq. (6-9)] at each value of m. Another useful aspect of the normal distribution is the area under the curve between two limits, which is the *cumulative distribution function* (CDF), the integral of the probability density function. Equation (6-11) shows the continuous integral

$$G(\sigma, \mu) = \frac{1}{\sqrt{2\pi}\sigma} \int_{\lambda_1}^{\lambda_2} \exp\left[-\frac{1}{2}\left\{\frac{\lambda - \mu}{\sigma}\right\}^2\right] d\lambda; \qquad \lambda_1 \leq m \leq \lambda_2$$

$$(6\text{-}11)$$

where λ is a *dummy variable* of integration. The value of this integral from $-\infty$ to $+\infty$ for finite values of σ and μ is exactly 1.0, which corresponds to 100% probability. Approximate values of this integral are available *only* in lookup tables or by various numerical methods. For a relatively easy method, use a favorite search engine to look up the "trapezoidal rule" or some other rule, or use programs such as Mathcad that have very sophisticated integration algorithms that can very quickly produce 1.0 ± 10^{-12} or better.

The area (CDF) for fractions of σ (called $x\sigma$) can be estimated visually using Fig. 6-3, where x is the variable of integration in the equation in Fig. 6-3 and the value of $\mu = 0$. The $G(x)$-axis value is the area (CDF) under the PDF curve from 0 to $x\sigma$, and the horizontal axis applies to values of $x\sigma$ from 0.01 to 3.0. The value of $x\sigma$ must be ≤ 3 for a good visual estimate. If $x\sigma = 0.50$, the area (CDF) from 0 to $0.5 \approx 0.19$. This graph is universal and applies to any σ value.

To get the total area (CDF) for a combination of $x\sigma > 0$ and $x\sigma < 0$, get the area $G(x\sigma)$ values between the boundaries of the $x\sigma > 0$ range. Use the positive region in the graph also to get the area for the $x\sigma < 0$ range and add the two positive-valued results (the normal PDF curve is symmetrical about the 0 value). The final sum should be no greater than $+1.0$.

The basic ideas in this section regarding the normal distribution apply with some modifications to other types of statistics, which can be explored in greater detail in the literature, e.g., [Meyer, 1970] and [Zwillinger, 1996, Chap. 7].

CORRELATION AND COVARIANCE

Correlation and covariance are interesting subjects that are very useful in noise-free and noise-contaminated electronic signals. They also lead to useful ideas in system analysis in Chapter 7. We can only touch briefly on these rather advanced subjects. Correlation is of two types: autocorrelation and cross-correlation.

Autocorrelation

In autocorrelation, a discrete-time sequence $x(n)$, with additive noise $\varepsilon(n)$, is sequence-multiplied (Chapter 5) by a time-shifted (τ) replica of itself. The discrete-time equation for the autocorrelation of a discrete-time sequence with noise ε is

$$C_A(\tau) = \frac{1}{N} \sum_{n=0}^{N-1} \left\{ [x + \varepsilon_x]_n \times [x + \varepsilon_x]_{(n+\tau)} \right\} \tag{6-12}$$

in which the integer τ is the value of the time shift from (n) to $(n + \tau)$. Each term $(x + \varepsilon_x)_n$ is one sample of a time sequence in which each has amplitude plus noise and time-position attributes.

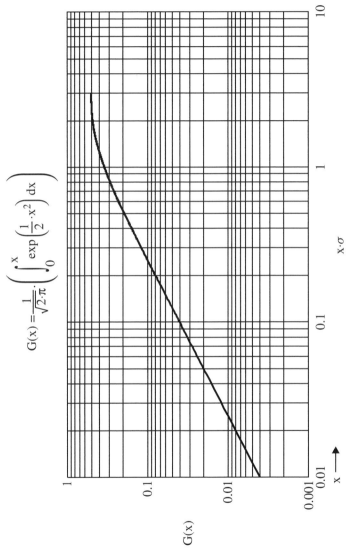

Figure 6-3 Probability CDF from $x\sigma = 0.01\sigma$ to 3.0σ for the normal distribution.

$$G(x) = \frac{1}{\sqrt{2\cdot\pi}}\cdot\left(\int_0^x \exp\left(\frac{1}{2}\cdot x^2\right) dx\right)$$

Equation (6-12) is assumed, as usual, to be one record of a steady-state repetitive sequence. Note that the "flip" of $x(n)$ does not occur as it did in Eq. (5-4) for convolution. We only want to compare the sequence with an exact time-shifted replica. Note also the division by N because $C_A(\tau)$ is by definition a time-averaged value for each τ and convolution is not. As such, it measures the average power *commonality* of the two sequences as a function of their separation in time. When the shift $\tau = 0$, $C_A(\tau) = C_A(0)$ and Eq. (6-12), reduces to Eq. (6-5), which is by definition the average power for $(x + \varepsilon_x)_n$.

Figure 6-4 is an example of the autocorrelation of a sequence in part (a) (no noise) and the identical shifted ($\tau = 13$) sequence in part b, c. There are three overlaps, and the values of the autocorrelation vs overlap, which is the sum of partial products (polynomial multiplication), are shown in part (c). The correlation value for $\tau = 13$ is

$$C_A(13) = \frac{(1)(0.1875) + (0.9375)(0.125) + (0.875)(0.0625)}{16} = 0.0225$$

This value is indicated in part (c), third from the left and also third from the right. This procedure is repeated for each value of τ. At $\tau = 0$, parts (a) and (b) are fully overlapping, and the value shown in part (c) is 0.365. For these two identical sequences, the maximum autocorrelation occurs at $\tau = 0$ and the value 0.365 is the average power in the sequence.

Compare Fig. 6-4 with Fig. 5-4 to see how circular autocorrelation is performed. We can also see that $x_1(n)$ and $x_2(n)$ have 16 positions and the autocorrelation sequence has $33 = (16 + 16 + 1)$ positions, which demonstrates the same smoothing and stretching effect in auto correlation that we saw in convolution. As we decided in Chapter 5, the extra effort in circular correlation is not usually necessary, and we can work around it.

Cross-Correlation

Two different waveforms can be completely or partially dependent or completely independent. In each of these cases the two noise-contaminated waveforms are time-shifted with respect to each other in increments of τ.

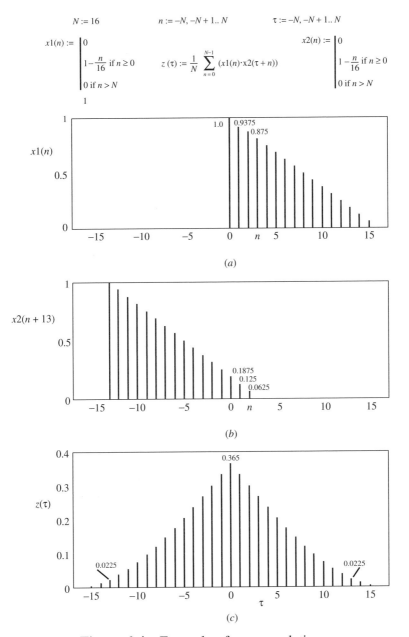

Figure 6-4 Example of autocorrelation.

Equation (6-13) is the basic equation for the cross-correlation of two different waves $x(n)$ and $y(n)$:

$$C_C(\tau) = \frac{1}{N} \sum_{n=0}^{N-1} \left[(x + \varepsilon_x)_n (y + \varepsilon_y)_{(n+\tau)} \right] \tag{6-13}$$

We have pointed out one major difference between the correlation and convolution equations. In correlation there is no "flip" of one of the waves, as explained in Chapter 7. This is in agreement with the desire to compare a wave with a time-shifted replica of itself or a replica of two different waves, one of which is time-shifted with respect to the other. In the case of convolution we derived a useful relationship for the Fourier transform of convolution. In Chapter 7, correlation leads to another useful idea in linear analysis, called the Wiener-Khintchine (see Google, e.g.) principle.

Figure 6-5 (with no noise) is an example of cross-correlation. The two time-domain sequences can have different lengths, different shapes, and different amplitude scale factors. The maximum value of cross-correlation occurs at $\tau = -3$ and -4, which is quite a bit different from Fig. 6-4. At $\tau = 0$ the correlation is 0.096, and at $\tau = -3$ and -4 the correlation is about 0.149, so the correlation in the overlap area increases 10 $\log(0.149/0.096) = 1.90$ dB. Recall that for each value of τ the area of overlap (sum of products as in Fig. 6-4) of the two sequences represents a value of *common power*. This value is the power that the two different waves deliver in combination.

The correlation *sequences* in Figs. 6-4 and 6-5 are τ-domain power sequences. These *power sequences* can also have complex (real watts and imaginary vars) frequency-domain components, just like any other time-domain sequence. The result is a power spectrum (Chapter 7) of correlation parameter τ.

AUTOCOVARIANCE

The calculation of autocorrelation can produce an average term, perhaps dc, which may not be useful or desired for statistical analysis reasons and should be eliminated. To accomplish this, autocovariance equation (6-14)

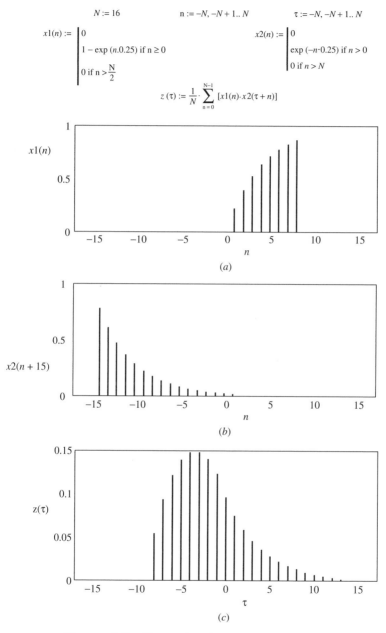

Figure 6-5   Example of cross-correlation.

removes the time average $\langle n_x \rangle$ and the result is a good approximation to the ac value expected. Many repetitions of Eq. (6-14), followed by averaging, can greatly improve the accuracy. This equation also leads to an ac energy or power result as a function of τ. If $\tau = 0$, the result is the average ac signal plus noise power in the $x(n)$ signal.

$$C_{acv}(\tau) = \frac{1}{N} \sum_{n=0}^{N-1} \left([(x + \varepsilon)_n - \langle n_x \rangle] \cdot [(x + \varepsilon)_{(n+\tau)} - \langle n_x \rangle] \right) \quad (6\text{-}14)$$

An example of autocovariance is the same as Fig. 6-4, which has been modified to remove the dc component.

Cross-Covariance

The same modification of the cross-correlation of two separate waves, $x(n)$ and $y(n)$, eliminating $\langle n_x \rangle$ and $\langle n_y \rangle$, produces the cross-covariance

$$C_{CCV}(\tau) = \frac{1}{N} \sum_{n=0}^{N-1} \left([(x + \varepsilon_x)n - \langle n_x \rangle] \cdot [(y + \varepsilon_y)_{(n+\tau)} - \langle n_y \rangle] \right) \quad (6\text{-}15)$$

The cross-covariance is the ac signal power plus noise power that is *common* to $x(n)$ and $y(n)$ as a function of shift τ. The result is the relatedness of the two ac signal powers. At any value of τ the result is the total ac power for that τ. Again, the result should be averaged over many repetitions.

Correlation Coefficient

This is an important dimensionless number in statistics, in addition to those just considered. Its value lies between -1.0 and $+1.0$ and it is a measure of the "relatedness" in some sense (to be decided by the user) between two possibly noise-contaminated sequences $x(n)$ and $y(n)$. The value -1.0 means "negatively" related, $+1.0$ means "positively" related, $|\rho_{xy}| = 1$ means completely related one way or the other, and 0 means that $x(n)$ and $y(n)$ are completely unrelated (independent). The basic equation for it is Eq. (6-16), where we use E (expectation) to mean the same as "averaging", assuming that many repetitions have been performed:

$$\rho_{xy} = \frac{E\{[x(n) - E(x(n))][y(n) - E(y(n))]\}}{\sqrt{V(x(n))V(y(n))}} \tag{6-16}$$

$$= \frac{E\{[x(n) - E(x(n))][y(n) - E(y(n))]\}}{\sigma_X \sigma_Y}$$

After many repetitions and averaging of ρ_{xy}, the numerator is the expected value of the cross-covariance of $x(n)$ and $y(n)$ [Eq. (6-15)], and the denominator is the square root of the product of the variances of $x(n)$ and $y(n)$, or more simply, just $\sigma_x \sigma_y$. $V(x(n))$ and $V(y(n))$ or (σ_x and σ_y) must both be greater than 0.0. This equation can be simplified as

$$\rho_{xy} = \frac{E[x(n)y(n)] - E[x(n)]E[y(n)]}{\sqrt{V(x(n))V(y(n))}} \tag{6-17}$$

$$= \frac{E[x(n)y(n)] - E[x(n)]E[y(n)]}{\sigma_X \sigma_Y}$$

If $x(n)$ and $y(n)$ are independent then the numerator of Eq. (6-17) is zero:

$$E[x(n)y(n)] = E[x(n)]E[y(n)] \tag{6-18}$$

and $\rho_{xy} = 0.0$. However, there are some cases, not to be explored here, where $x(n)$ and $y(n)$ are not independent, yet ρ_{xy} is nevertheless equal to zero. So "uncorrelated" and "independent" do not always coincide. Looking at Eq. (6-18), we can guess that this might happen. For further insight about the correlation coefficient, see [Meyer, 1970, Chap. 7].

As an example we will calculate ρ_{xy} of Fig. 6-5 using Eq. (6-17) and the time-averaged values instead of expected values because Eq. (6-17) is assumed to be noise-free:

$$\rho_{XY} = \frac{\langle(xy)\rangle - \langle x\rangle\langle y\rangle}{\sigma_X \sigma_Y} \tag{6-19}$$

$$= \frac{0.096 - (0.31 \cdot 0.277)}{0.344 \cdot 0.286} = 0.099$$

The same calculation on Fig. 6-4 produces a value of 1.00.

This brief introduction to correlation and variance is no more than a "get acquainted" starting point for these topics and is not intended as a substitute for more advanced study and experience with probability and statistical methods. We are limited to signal sequences that are discrete in both time and frequency domains from 0 to $N - 1$, which makes things a little easier. Mathcad calculates very easily all of the equations in this chapter.

REFERENCES

Carlson, A. B., 1986, *Communication Systems*, 3rd ed., McGraw-Hill, New York.

Meyer, P. L., 1970, *Introductory Probability and Statistical Methods*, Addison-Wesley, Reading, MA.

Oppenheim, A. V., and R. W. Schafer, 1999, *Discrete-Time Signal Processing*, 2nd ed., Prentice Hall, Upper Saddle River, NJ.

Schwartz, M., 1980, *Information Transmission, Modulation and Noise*, 3rd ed., McGraw-Hill, New York.

Zwillinger, D., Ed., 1996, *CRC Standard Mathematical Tables and Formulae*, CRC Press, Boca Raton, FL.

7

The Power Spectrum

We have learned that a time-domain discrete sequence $x(n)$ that extends from $0 \le n \le N - 1$ can be considered as two-sided, positive-time for the first half and negative-time for the second half. Each sample $x(n)$, considered by itself, is just a *magnitude* (see Chapter 1). It also has a time-position attribute but none other, such as frequency or phase or properties such as real or imaginary. In other words, $x(n)$ is not a phasor. It is what we see on an ordinary oscilloscope.

On the other hand, the $x(n)$ *sequence* (the entire scope screen display) can consist of a set of complex-valued voltage or current *waveforms* applied to a complex load network of some kind. However, time-domain analysis of complex signals combined with complex loads requires math methods that we will not explore in this book [Oppenheim and Schafer, 1999; Carlson, 1986; Schwartz, 1980; Dorf and Bishop, 2005; Shearer et al., 1971], so we prefer to convert the time sequence $x(n)$ to the frequency $X(k)$ domain using the DFT. After processing the signal in the frequency domain we can, if we wish, use the IDFT to get the time domain $x(n)$ sequence representation of the processed discrete signal.

Discrete-Signal Analysis and Design, By William E. Sabin
Copyright © 2008 John Wiley & Sons, Inc.

This is a simple and very useful approach that is widely used, especially in computer-aided design.

In this chapter we are interested in power. We are also interested in phasors. The problem is that any phasor that has constant amplitude has zero average power, so it makes no sense to talk about average phasor power. Therefore, we will combine the positive- and negative-frequency phasors coherently, using the methods described in Fig. 2-2 and employed elsewhere, to get a positive-frequency sine wave or cosine wave at frequency (k) and phase $\theta(k)$ from $1 \le k \le N/2 - 1$. We then have a true signal that has average power at frequency (k), and we can look at its power spectrum.

There is another approach available. The real or imaginary part of the phasor $Me^{j\omega t}$ is a sinusoidal wave that has a peak value M. The rms value of this sinusoidal wave, considered by itself, is $M \cdot 0.7071$. In our Mathcad examples the method of the previous paragraph, where we combine both sides of the phasor spectrum coherently, is an excellent and very simple approach that takes into account the two complex-conjugate phasors that are the constituents of the true sine or cosine signal.

FINDING THE POWER SPECTRUM

We will use voltage values, but current values apply equally well, using the Norton source transformation [Shearer et al., 1971]. The discrete Fourier transform DFT [Eq. (1-2)] of an $x(n)$ signal *sequence* leads to a discrete two-sided $X(k)$ steady-state spectrum of complex voltage phasors oscillating at frequency (k) from 1 to $N - 1$ with amplitude $X(k)$ and relative phase $\theta(k)$. At each (k) and $(N - k)$ we will combine a pair of complex-conjugate phasors to get a positive-side sine or cosine $V(k)$ from $k = 1$ to $k = N/2 - 1$.

In order to keep the analysis consistent with circuit realities, assume that $V(k)$ is an open-circuit voltage generator whose internal impedance is for now a constant resistance R_g, but a complex $Z_g(k)$ can very easily be used. $V(k)$ is then the steady-state open-circuit voltage at frequency (k). The voltage $VL(k)$ across the load $Y(k)$ (see Fig. 7-1) is found independently

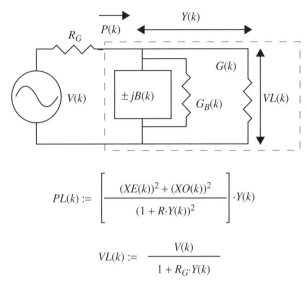

$$PL(k) := \left[\frac{(XE(k))^2 + (XO(k))^2}{(1 + R \cdot Y(k))^2} \right] \cdot Y(k)$$

$$VL(k) := \frac{V(k)}{1 + R_G \cdot Y(k)}$$

Figure 7-1 Equivalent circuit of frequency-domain power spectrum at frequency position k.

at each frequency (k) as

$$VL(k) = \frac{V(k)}{1 + R_G Y(k)} = \frac{V(k)}{1 + Z_G(k)Y(k)} \tag{7-1}$$

For a given $V(k)$ the steady-state value of $VL(k)$ depends only on the present value of $Y(k)$ and $Z_G(k)$ or R_G, and not on previous or future values of (k). The complex load admittance $Y(k) = [G(k) + G_B(k)] \pm jB(k)$ siemens which we have pre-determined by calculation or measurement at each frequency (k), is driven by the complex signal voltage $VL(k) = \mathrm{Re}[VL(k)] \pm j\mathrm{Im}[VL(k)]$, and the complex power to the load is

$$\mathrm{PL}(k) = VL(k)^2 Y(k) \quad \text{watts and vars} \tag{7-2}$$

Figure 7-1 shows the equivalent circuit for Eq. (7-1). $PL(k)$ is the power spectrum with real part (watts), imaginary part (vars), and phase angle $\theta(k)$, that is delivered to the complex load admittance $Y(k) = G(k) + G_B(k) \pm jB(k)$. If $B(k)$ is zero, the power $Pl(k)$ is in phase with

$VL(k)$, $PL(k) = VL(k)^2[G_B(k) + G(k)]$ watts. If $G(k) + G_B(k) = 0$, the imaginary power (vars) $PL(k) = \pm jB(k) \times VL(k)^2$.

The real part of the power $PL(k)$ is converted to radio or sound waves or heat dissipation of some kind, and the imaginary part is cycled back and forth between energy storage elements (lumped components) or standing waves (transmission lines) of some kind. This energy cycling always involves slightly lossy storage elements that dissipate a little of the real power.

Example 7-1: The Use of Eq. (7-2)

Figure 7-2 is an example of the use of Eq. (7-2). The $x(n)$ input signal voltage waveform in part (a) is a complex time sequence of cosine and sine waves. This figure uses steps of 0.1 in the (n) values for better visual resolution, and this is the only place where $x(n)$ is plotted. The two plots in part (a) are $I(n)$ (real) and $Q(n)$ (imaginary) sequences that we have looked at previously. Parts (b) and (c) are the DFT of part (a) that show the two-sided phasor frequency $X(k)$ voltage values. The DFT uses (k) steps of 1.0 to avoid spectral leakage between (k) integers (Chapter 3). If a dc voltage is present, it shows up at $k = 0$ (see Fig. 1-2). In this example there is no dc, but it will be considered later. The integer values are sufficient for a correct evaluation if there are enough of them to satisfy the requirements for adequate sampling.

In part (d) the two-sided phasors are organized into two groups. One group collects phasor pairs that have *even* symmetry about $N/2$ and are added coherently (Chapter 1). These are the cosine (or j cosine) terms. The other pairs that have *odd* symmetry about $N/2$ are the sine (or the j sine) terms and are subtracted coherently. This procedure accounts for all phasor pairs in any signal, regardless of its even and odd components, and the results agree with Fig. 2-2. Plots (f) and (g) need only the positive frequencies. Note also that the frequency plots are not functions of time, like $x(n)$, so each observation at frequency (k) is a steady-state measurement and we can take as much time at each (k) as we like, after the $x(n)$ time sequence is obtained.

Part (e) calculates the load admittance $Y(k) = G(k) \pm jB(k)$ at each (k) for the frequency dependence that we have specified. The plot in part (f) shows the complex value of $Y(k)$ at each (k).

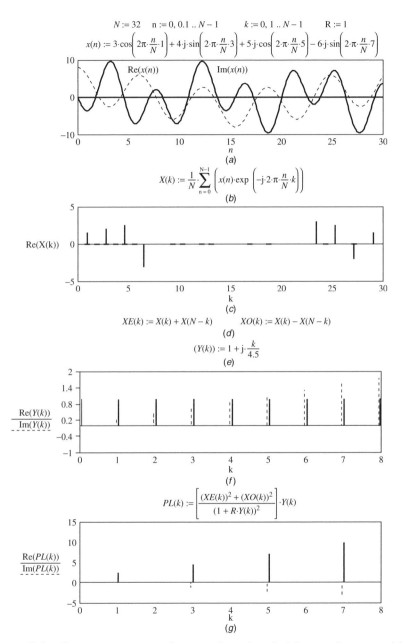

Figure 7-2 Power spectrum of a complex signal: (a) complex two-sided time domain, real and imaginary; (b) complex two-sided phasor voltage spectrum; (c) complex two-sided phasor voltage spectrum; (d) even and odd parts of phasor spectrum; (e) load admittance definition; (f) load admittance plot; (g) load power spectrum after filtering.

In part (g), Eq. (7-2), the *positive-frequency* complex power $PL(k)$ as influenced by the complex load admittance $Y(k)$, is calculated and plotted. This power value is due to two separate and independent power contributions. The first is due only to the $XE(k)$ terms (the even terms) that are symmetrical about $N/2$. The second is due only to the $XO(k)$ terms (the odd terms) that are odd-symmetrical about $N/2$. In other words, the power spectrum is a linear collection of sinusoidal power signals, which is what the Fourier series is all about. If a certain $PL(k)$ has a phase angle associated with it, Mathcad separates $PL(k)$ into an even (real) part and an odd (imaginary) part. Also, the power spectrum can have an imaginary part that is negative, which is determined in this example by the definition of $Y(k)$, and the real part is positive in passive networks but can have a negative component in amplifying feedback networks [Gonzalez, 1997].

If there is a dc component in the input signal $x(n)$ in part (a), a dc voltage will be seen in part (c) at zero frequency. Part (g) includes this dc voltage in its calculation of the dc power component in $PL(k)$. Note that part (f) shows an admittance value at $k = 0$ that can be forced to zero using a dc block (coupling capacitor or shunt inductor).

As an alternative to the use of exact-integer values of (k) and (n), the windowing and smoothing procedures of Chapter 4 can greatly reduce the spectral leakage sidelobes (Chapter 3), and that is a useful approach in practical situations where almost-exact-integer values of (n) and (k) using the rectangular window are not feasible. The methods in Chapter 4 can also reduce aliasing (Chapter 3). These windowing and smoothing functions can easily be appended to $x(n)$ in Fig. 7-2a.

At this point we would like to look more closely at random noise.

RANDOM GAUSSIAN NOISE

The product of temperature T in Kelvins and k_B, Boltzmann's constant $(1.38 \times 10^{-23}$ joules per kelvin), equals energy in joules, and in conducting or radiating systems this amount of energy flow per second at constant (or varying) temperature is $k_B T$ watts (joules per kelvin per second). T is quite often 290K or 17°C (63°F) and $k_B T = 4 \times 10^{-21}$ watts, the electrical *thermal noise power* that is available from any purely resistive electrical thermal noise source in a 1.0-Hz bandwidth at a chilly "laboratory" temperature, and

$$P(\text{avail}) = 10 \log \left[\frac{4.0 \times 10^{-21}}{0.001} (1.0) \right] = -174\,\text{dBm} \qquad (7\text{-}3a)$$

where the 0.001 converts watts to milliwatts. If the noise bandwidth is B and the temperature is T, then

$$P(\text{avail}) = \left[-174 + 10 \log(B/1.0) + 10 \log(T/290) \right] \text{dBm} \quad (7\text{-}3b)$$

Some resistance values are not sources of thermal noise and do not dissipate power. These are called *dynamic resistances*. One example is the lossless transmission line whose characteristic resistance, $R_0 = V_{ac}/I_{ac}$, such as 50 ohms. Another dynamic resistance is the plate (collector) resistance of a vacuum tube (transistor), dv/di, which is due to a lossless internal negative feedback effect. Also, a pure reactance is not a source of thermal noise power because the across-voltage and through-current are in phase quadrature (average power $= 0$).

This thermal noise has an inherent bandwidth of, not infinity, but up to about 1000 GHz, (1.0 THz) where quantum-mechanical effects involving Planck's energy constant (6.63×10^{-34} joule-seconds) start to cause a roll-off [Carlson, 1986, pp. 171–172]. For frequencies below this, Eq. (7-3) is the thermal noise available power spectral density unless additional filtering of some type further modifies it. We note also that Eq. (7-3) is the one-sided ($f \geq 0$) spectrum, 3 dB greater than the two-sided value. The two-sided value is sometimes preferred in math analyses. Thermal noise has the Gaussian (normal) probability density (PDF) and cumulative distribution (CDF) previously discussed in Chapter 6. The thermal noise signal is very important in low-level system simulations and analyses.

The term *white noise* refers to a constant wideband (<1000 GHz, 1 THz) value of power spectrum (see the next topic) and to essentially zero autocorrelation for $\tau \neq 0$ [Eq. (6-12)]. As the system bandwidth is greatly reduced, these assumptions begin to deteriorate. At narrow bandwidths an approach called *narrowband noise analysis* is needed, and real-world envelope detection of noise combined with a weak signal imposes additional nonlinear complications, including a "threshold" effect [Schwartz, 1980, Chap. 5; Sabin, 1988]. A common experience is that as the signal increases slowly from zero, an increase in noise level is noticed. At higher signal levels the noise level is reduced as the detector becomes more "linearized".

MEASURING POWER SPECTRUM

The spectrum analyzer (or scanning spectrum analyzer) is an excellent example of a power spectrum instrument. The horizontal scale (usually, a linear scale) indicates frequency increments, and the vertical scale is calibrated in dB with respect to some selected reference level in dBm at the top of the display screen. Frequency resolution bandwidth values from 1 Hz (very expensive) to 100 MHz are common. The data in modern instruments is stored digitally in one or more memory registers for each resolution bandwidth value, and this data can be processed in many different ways. We can think of the spectrum analyzer as a discrete-frequency sampler at frequencies (k). The amplitude $X(k)$ is usually also stored in digital form.

One especially interesting usage is the "peak hold" option, where the frequency range is scanned slowly several dozen times, and any increase in the peak value at any frequency is preserved and updated on each new scan. A steady-state pattern slowly emerges on the screen that is an accurate display of the power spectrum. The analysis of random or pseudorandom RF signal power spectra such as speech or data is greatly facilitated. The input signal must be strong enough to override external and internal noise contamination. Moderately priced instruments often have this very valuable feature, and this is an excellent way to evaluate signals that have long-term randomness.

Figure 7-3 is derived from a photograph of a spectrum analyzer display of a long-term peak hold of a radio frequency 600 watt PEP single-sideband adult male voice signal that shows a speech frequency passband from about 300 Hz to about 3 kHz above the suppressed carrier frequency f_0. The resolution bandwidth is 300 Hz. The lower speech frequencies are attenuated somewhat in order to emphasize the higher speech frequencies that improve readability under weak signal conditions. The low levels of adjacent channel spillover caused by the inevitable nonlinearities in the system (the transmitter) are also shown. Note the > 40-dB attenuation at

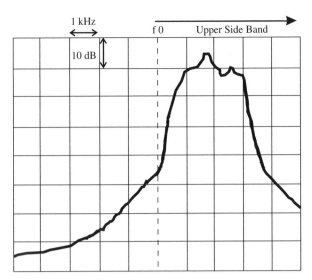

Figure 7-3 Single-sideband speech power spectrum, spectrum analyzer plot.

$(f_0 - 1)$ kHz and $(f_0 + 4)$ kHz. This kind of display would be difficult to obtain using purely mathematical methods because the long-term spectral components on adjacent channels caused by various mild system non-linearities combined with a very complicated complex signal would be difficult, but not impossible, to model accurately.

Another instrument, the vector network analyzer, displays dB amplitude and phase degrees or complex S-parameters in a polar or Smith chart pattern, which adds greatly to the versatility in RF circuit design and analysis applications. The important thing is that the signal is sampled in certain fixed and known bandwidths, and further analyses of the types that we have been studying, such as filtering, smoothing and windowing and others, both linear and nonlinear, can be performed on the data after it has been transferred from the instrument. This processed spectrum information can be transformed to the time or frequency domain for further evaluations.

Wiener-Khintchine Theorem

Another way to get a two-sided power spectrum sequence is to carry out the following procedures:

1. From the $x(n)$ time sequence, calculate the autocorrelation function $C_A(\tau)$ using Eq. (6-12). Note that τ is the integer value (0 to $N-1$) of shift of $x(n)$ that is used to get $C_A(\tau)$.

2. Perform the DFT on $C_A(\tau)$ using Eq. (1-2) to get $P(k)$ [Carlson, 1986, Sec. 3]. Note that the shift of τ is carried out in steps of 1.0 over the range from 0 to $N-1$ in Eq. (7-4).

$$P(k) = \mathcal{F}[C_A(\tau)] = \frac{1}{N}\sum_{\tau=0}^{N-1} C_A(\tau)\exp\left(-j2\pi\frac{\tau}{N}k\right) \qquad (7\text{-}4)$$

This $P(k)$ spectrum is two-sided and can be converted to one-sided as explained in Chapter 2 and earlier in this chapter. The Wiener-Khintchine theorem is bi-directional and the two-sided autocorrelation $C_A(\tau)$ can be found by performing the IDFT [Eq. (1-8)] on the two-sided $P(k)$:

$$C_A(\tau) = \mathcal{F}^{-1}[P(k)] = \sum_{k=0}^{N-1} P(k)\,\exp\left(j2\pi\frac{\tau}{N}k\right) \qquad (7\text{-}5)$$

The FFT can be used to expedite the forward and reverse Fourier transformations. This method is also useful for sequences that are unlimited (not periodic) in the time domain, if the autocorrelation function is available.

SYSTEM POWER TRANSFER

The autocorrelation and cross-correlation functions can be defined in terms of periodic repeating signals, in terms of finite nonrepeating signals, and in terms of random signals that may be infinite and nonrepeating [Oppenheim and Schafer, 1999, Chap. 10].

We have said that for this introductory book we will assume that a sequence of 0 to $N-1$ of some reasonable length N contains enough significant information that all three types can be calculated to a sufficient degree of accuracy using Eqs. (6-12) and (6-13). We will make N large enough that circular correlation and circular convolution are not needed. We will continue to assume an infinitely repeating process. When a fairly low value of noise contamination is present, we will perform averaging of many sequences to get an improved estimate of the correct values. We will also assume ergodic, wide-sense stationary processes that make our assumptions reasonable. This means that expected (ensemble) value and time average are "nearly" equal, especially for Gaussian noise. We also assume that windowing and anti-aliasing procedures as explained in Chapters 3 and 4 have been applied to keep the 0 to $N-1$ sequence essentially "disconnected" from adjacent sequences. The Hanning and Hamming windows are especially good for this.

If a linear system, possibly a lossy and complex network, has the complex voltage or current input-to-output frequency response $H(k)$ and if the input power spectrum is $P(k)_{in}$, the output power spectrum $P(k)_{out}$ in a 1.0 ohm resistor can be found using Eq. (7-6)

$$P(k)_{out} = [H(k)H(k)^*]P(k)_{in} = |H(k)|^2 P(k)_{in} \qquad (7\text{-}6)$$

where the asterisk(*) means complex conjugate. Because $P(k)_{in}$ and $P(k)_{out}$ are Fourier transforms of an autocorrelation, their values are real and nonnegative and can be two-sided in frequency [Papoulis, 1965, p. 338]. This is an important fundamental idea in the design and analysis of linear systems. Equation (7-6) is related to the Fourier transform of convolution that we studied and verified in Eqs. (5-6) to (5-10). Equation (7-6) for the power domain is easily deduced from that material by including the complex conjugate of $H(k)$. To repeat, $P(k)_{in}$ and $P(k)_{out}$ are real-valued, equal to or greater than zero and two-sided in frequency.

CROSS POWER SPECTRUM

Equation (7-4) showed how to use the auto-correlation in Eq. (6-12) to find the power spectrum of a single signal using the DFT. In a similar manner, the cross-spectrum between two signals can be found from the DFT of the cross-correlation in Eq. (6-13). The cross-spectrum evaluates the *commonality* of the power in signals 1 and 2, and phase commonality is included in the definition. We will now use an example of a pair of sinusoidal signals to illustrate some interesting ideas.

Equation (7-7) compares the average power P_1 for the product of a sine wave and a cosine wave on the same frequency, and the average power P_2 in a single sine wave. P_3 is the average power for the sum of two sine waves in phase on the same frequency. For better visual clarity we temporarily use integrals instead of the usual discrete summation formulas:

$$P_1 = \frac{1}{2\pi} \int_0^{2\pi} A \cos\theta B \sin\theta d\theta = \frac{AB}{2\pi} \int_0^{2\pi} \cos\theta \cdot \sin\theta d\theta = 0$$

$$P_2 = \frac{1}{2\pi} \int_0^{2\pi} (\sin\theta)^2 d\theta = \frac{1}{2\pi} \cdot \int_0^{2\pi} \sin\theta \sin\theta d\theta = 0.5 \qquad (7\text{-}7)$$

$$P_3 = \frac{1}{2\pi} \int_0^{2\pi} (\sin\theta + \sin\theta)^2 d\theta = \frac{1}{2\pi} \int_0^{2\pi} 4(\sin\theta)^2 d\theta = 2.0$$

The trig identities confirm the values of the *integrals* for P_1, P_2 and P_3. In P_1 the two are $90°$ out of phase and the integral evaluates to zero. Note that P_1 (only) is zero for any real or complex amplitudes A and B. However, a very large product $A \cdot B$ can make it difficult to make the numerical integration of the product $(\cos\theta) \cdot (\sin\theta)$ actually become very small. To repeat, P_2 is the average power of a single sine wave.

We can also compare P_1 and P_2 using the cross-correlation Eq. (6-13). P_2 is the product of two sine waves with $\tau = 0$. The cross-correlation, and therefore the cross power spectrum, is maximum. P_1 is the cross-correlation of two sine waves with $\tau = \pm 1/4$ cycle applied to the left-hand sine wave. The cross-correlation is then zero and the cross power spectrum is also zero, applying the Wiener-Khintchine theorem to Eq. (6-13).

In P_3 the two are $0°$ in-phase (completely correlated) and the sum of two sine waves produces an average power of 2.0, four times (6 dB greater than) the average power P_2 for a single sine wave. If the two waves in P_3 were on greatly different frequencies, in other words uncorrelated, each would have an average power of 0.5 and the total average power would be 1.0. This means that linear superposition of independent (uncorrelated) power values can occur in a linear system, but if the two waves are identically in phase, an additional 3 dB is achieved. The generator must deliver 3 dB more power. P_3 for the sum of a sine wave and a cosine wave $= 0.5 + 0.5 = 1.0$ because the sine and cosine are independent (uncorrelated). Also, inside a narrow passband the correlation (auto or cross) value does not suddenly go to zero for slightly different frequencies; instead, it decreases smoothly from its maximum value at

$\Delta f = 0$, and more gradually than in a wider passband [Schwartz, 1980, p. 471]. Coherence is used to compare the relationship, including the phase relationship, of two sources. If they are all fully in phase, they are fully coherent. Coherence can also apply to a constant value of phase difference. The coherence number ρ between spectrum power S_1 and spectrum power S_2 can be found from Eq. (7-8).

$$\rho = \frac{\text{cross power spectrum}}{\sqrt{S_1 S_2}}, \quad \rho \leq 1.0 \qquad (7\text{-}8)$$

Finally, two independent uncorrelated signals in the same frequency passband, each with power 0.5, produce a peak envelope power (PEP) $= 2.0$ (6 dB greater) and an average power $= 1.0$ (3 dB greater) [Sabin and Schoenike, 1998, Chap. 1]. The system must deliver this PEP with low levels of distortion.

As we said before [Eq. (7-7)], if two pure sinusoidal signals at the same amplitude and frequency are 90 degrees out of phase, the average power in their product is zero. But if these signals are contaminated with *amplitude noise*, or often more important, *phase noise*, the two signals do not completely cancel. The combination of phase noise and amplitude noise is known as *composite noise*. The noise spectrum can have a bandwidth that degrades the performance of a phase-sensitive system or some adjacent channel equipment.

Measurement equipment that compares one relatively pure sine wave and a test signal that is much less pure is used to quantify the noise contamination and spectrum of the test signal. It is also possible to compare two identical sources and calculate the phase noise of each source. The $90°$ phase shift that greatly attenuates the product at baseband of the two large sine-wave signals is important because it allows the residual unattenuated phase noise to be greatly amplified for easier measurement. A lowpass filter attenuates each input tone and all harmonics. A great deal of interest and effort are directed to tests of this kind and some elegant test equipment is commonly used.

Example 7-2: Calculating Phase Noise

An example of phase noise is shown in Fig. 7-4. What follows is a step-by-step description of the math. This is also an interesting example of discrete-signal analysis.

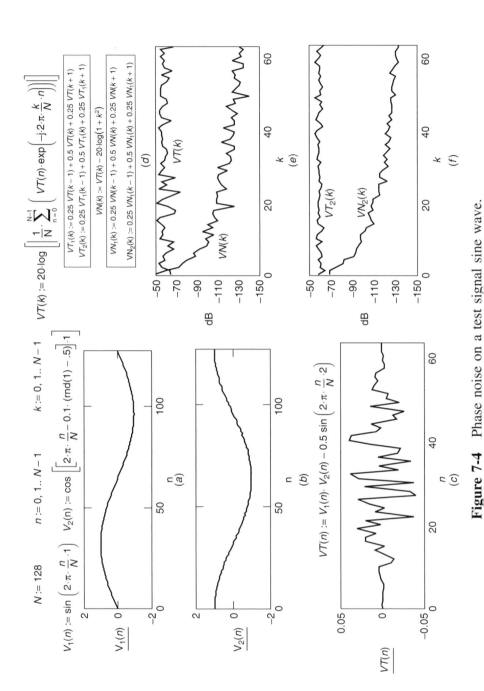

Figure 7-4 Phase noise on a test signal sine wave.

At the upper left is a noise-free discrete sine wave $V_1(n)$ at frequency $f = 1.0$, amplitude 1.0, in 128 positions of (n). A discrete cosine wave $V_2(n)$, amplitude 1.0 at the same frequency, has some phase noise added, $0.1 \cdot [\text{rnd}(1) - 0.5]$. The rnd(1) function creates a random number from 0 to $+1$ at each position of (n). The value 0.5 is subtracted, so the random number is then between -0.5 and $+0.5$. The index of the phase modulation is 0.1.

(a) The plot of the noise-free sine wave.

(b) The plot of the cosine wave with the noise just barely visible.

(c) We now multiply the sine wave and the cosine wave. This multiplication produces the baseband phase noise output $VT(n)$ and a sine wave of amplitude 0.5 at twice the frequency of the two input waves. We subtract this unwanted wave so that only the phase noise is visible in part (c). This is equivalent to a lowpass filter that rejects the times 2 frequency. Note the vertical scale in the graph of part (c) that shows the phase noise greatly amplified.

(d) We next use the DFT to get the noise spectrum $VT(k)$ in dB format. At this point we also perform two 3-point smoothing operations on $VT(k)$, first to get $VT_1(k)$ and then to get $VT_2(k)$. This operation smoothes the spectrum of $VT(k)$ so that $VT(k)$ in the graph in part (e) is smoothed to $VT_2(k)$ in the graph in part (f). This is *postdetection filtering* that is used in spectrum analyzers and many other applications to get a smoother appearance and reduce noise peaks; it improves "readability" of the noise shelf value.

(e) Also in part (d) we perform lowpass filtering $[-20 \log(1 + k^2)]$ (a Butterworth lowpass filter) to get $VN(k)$. This result is also smoothed two times and the comparison of $VN(k)$ and $VN_2(k)$ is seen in the graphs of parts (e) and (f).

(f) Note that in parts (e) and (f) the upper level of the phase noise plot $VT(k)$ and $VT_2(k)$ is >53 dB below the 0-dB reference level of the test signal V_2 at frequency $k = 1$. This is called the *relative noise shelf* for the noisy test signal that we used. It is usually expressed as a dBc number (dB below the carrier, in this case, >53 dBc). This noise shelf is of great significance in equipment design. It defines the ability to reject interference to and from closely adjacent signals and also to analyze unwanted phase disturbances on input signals. The lowpass filter to get $VN_2(k)$ greatly improves phase noise, but only at frequencies somewhat removed from the signal frequency. Still, it is very important that wideband phase noise interference is greatly reduced.

In conclusion, there are many advanced applications of the cross power spectrum that we cannot cover in this book but that can be explored using various search engines and texts.

REFERENCES

Carlson, A., 1986, *Communication Systems*, 3rd ed., McGraw-Hill, New York.

Dorf, R. C., and R. H. Bishop, 2005, *Modern Control Systems*, 10th ed., Prentice Hall, Englewood Cliffs, NJ.

Gonzalez, G., 1997, *Microwave Transistor Amplifier Analysis and Design*, Prentice Hall, Upper Saddle River, NJ.

Oppenheim, A. V., and R. W. Schafer, 1999, *Discrete-Time Signal Processing*, 2nd ed., Prentice-Hall, Upper Saddle River, NJ, p. 189.

Papoulis, A., 1965, *Probability, Random Variables, and Stochastic Processes*, McGraw-Hill, New York.

Sabin, W. E., 1988, Envelope detection and noise figure measurement, *RF Design*, Nov., p. 29.

Sabin, W. E., and E. O. Schoenike, 1998, *HF Radio Systems and Circuits*, SciTech, Mendham, NJ.

Schwartz, M., 1980, *Information Transmission, Modulation and Noise*, 3rd ed., McGraw-Hill, New York, Chap. 5.

Shearer, J. L., A. T. Murphy, and H. H. Richardson, 1971, *Introduction to System Dynamics*, Addison-Wesley, Reading, MA.

8

The Hilbert Transform

D. Hilbert 1862-1943

This final chapter considers a valuable resource, the Hilbert transform (HT), which is used in signal-processing systems to achieve certain properties in the time domain and the frequency domain. The DFT, IDFT, FFT, and Hilbert transform work quite well together with discrete signals if certain problem areas to be discussed are handled correctly.

Example of the Hilbert Transform

Figure 8-1 shows a two-sided square wave $x(n)$ time sequence, and we will walk through the creation of an HT for this wave. Design the two-sided square-wave time sequence using $N = 128$. The value at $n = 0$ is zero, which provides a sloping leading edge for better plot results. Values from 1 to $N/2 - 1 = +1.0$. Set the $N/2$ position to zero, which has been found to be important for successful execution of the HT because $N/2$ is a special location that can cause problems because of its small but non-zero value. Values from $N/2 + 1$ *to* $N - 1 = -1$. At N the wave returns to zero.

(a) Plot the two-sided square wave from $n = 0$ to $N - 1$.

Discrete-Signal Analysis and Design, By William E. Sabin
Copyright © 2008 John Wiley & Sons, Inc.

(b) Execute the DFT to get $X(k)$, the two-sided, positive-frequency first-half and negative-frequency second-half phasor spectrum. This spectrum is a set of sine waves as shown in Fig. 2-2c. Each sine wave consists of two imaginary components.

(c) Multiply the first-half positive-frequency $X(k)$ values by $-j$ to get a $90°$ phase lag. Multiply the second-half negative-frequency $X(k)$ values by $+j$ to get a $90°$ phase lead. As in step (a), be sure to set the $N/2$ value to zero. This value can sometimes confuse the computer unless it is forced to zero. This step (c) is the ideal Hilbert transform.

(d) Part (d) of the graph shows the two-sided spectrum $XH(k)$ after the phase shifts of part (c). This spectrum is a set of negative cosine waves as shown in Fig. 2-2b. Each cosine wave consists of two real negative components, one at $+k$ and one at $N-k$.

(e) Use the IDFT to get the two-sided $xh(n)$ time response. Note that $xh(n)$ is a sequence of real numbers because $x(n)$ is real.

(f) The original square wave and its HT are both shown in the graph. Note the large peaks at the ends and in the center. These peaks are characteristic of the HT of an *almost* square wave. The perfect square wave would have infinite (very undesirable) peaks.

(g) Calculate two 3-point smoothing sequences described in Chapter 4 for the sequence in part (f). This smoothing is equivalent to a lowpass filter, or at radio frequencies to a narrow band-pass filter.

(h) Plot the final result. The sharp peaks have been reduced by about $2\,\text{dB}$. Further smoothing is usually required in narrowband circuit design applications, as we will see later.

The three peaks in part (f) are usually a problem in any peak-power-limited system (which is almost always the practical situation). The smoothing in part (h) thus becomes important. Despite the peaks, the rms voltage in part (f) is the same for both of the waveforms in that diagram (nothing is lost).

For problems of this type, the calculation effort becomes extensive, and the use of the FFT algorithm, with its greater speed, would ordinarily be preferred. The methods of the Mathcad FFT and IFFT functions are described in the *User Guide* and especially in the online Help. In this

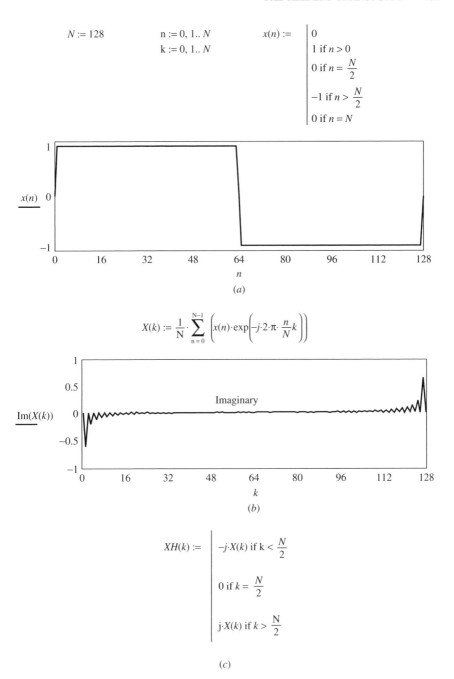

Figure 8-1   Example of the Hilbert transform.

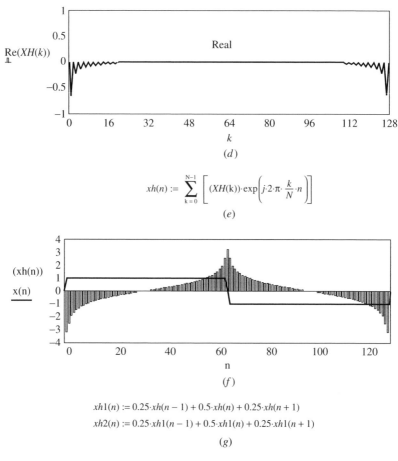

$$xh(n) := \sum_{k=0}^{N-1} \left[(XH(k)) \cdot \exp\left(j \cdot 2 \cdot \pi \cdot \frac{k}{N} \cdot n \right) \right]$$

(e)

(f)

$$xh1(n) := 0.25 \cdot xh(n-1) + 0.5 \cdot xh(n) + 0.25 \cdot xh(n+1)$$
$$xh2(n) := 0.25 \cdot xh1(n-1) + 0.5 \cdot xh1(n) + 0.25 \cdot xh1(n+1)$$

(g)

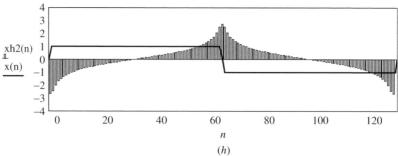

(h)

Figure 8-1   (*continued*)

chapter we will continue to use DFT and IDFT and stay focused on the main objective, understanding the Hilbert transform.

Why do the samples in Fig. 8-1f and h bunch up at the two ends and in the center to produce the large peaks? The answer can be seen by comparing Fig. 8-1b and d. In Fig. 8-1b we see a collection of (sine) wave harmonics as defined in Fig. 2-2c. These sine wave harmonics are the Fourier series constituents of the symmetrical square wave in Fig. 8-1a. In Fig. 8-1d we see a collection of (− cosine) waves as defined in Fig. 2-2b. These (− cosine) wave harmonic amplitudes *accumulate* at the endpoints and the center exactly as Fig. 8-1f and h verify. As the harmonics are attenuated, the peaks are softened. The smoothing also tends to equalize adjacent amplitudes slightly. The peaks in Fig. 8-1h rise about 8 dB above the square-wave amplitude, which is almost always too much. There are various ways to deal with this. One factor is that the square-wave input is unusually abrupt at the ends and center. Smoothing (equivalent to lowpass filtering) of the *input* signal $x(n)$, is a very useful approach as described in Chapter 4. This method is usually preferred in circuit design.

It is useful to keep in mind, especially when working with the HT, that the *quadrature* of θ, which is $θ \pm 90°$, is not always the same thing as the *conjugate* of θ. If the angle is $+30°$, its conjugate is $-30°$, but its quadrature is $+30° \pm 90° = +120°$ or $-60°$. The HT uses $θ \pm 90°$. For example, Fig. 8-1b shows $+j0.5$ *at* $k = 127$. The HT multiplies this by $+j$ to get a real value of -0.5 in part (d) at $k = 127$. This is a quadrature positive phase shift at negative frequency $k = 127$. A similar event occurs at $k = 1$. Also, at $k = 0$ and N the phase jump is $180°$ from $\pm j$ to $\mp j$, and the same, although barely noticeable, at $N/2$. (Use a highly magnified vertical scale in Mathcad.)

A different example is shown in Fig. 8-2. The baseband signal in part (b) is triangular in shape, and this makes a difference. The abrupt changes in the square wave are gone, the baseband spectrum (d) contains only real cosine-wave harmonics, and the Hilbert transform (f) contains only sine-wave harmonics. The sharp peaks in $xh(n)$ that we see in Fig. 8-1g disappear in Fig. 8-1h. However, for equal peak-to-peak amplitude, the square wave of Fig. 8-1 has 4.8 dB more average power than the triangular wave of Fig. 8-2. It is not clear what practical advantage

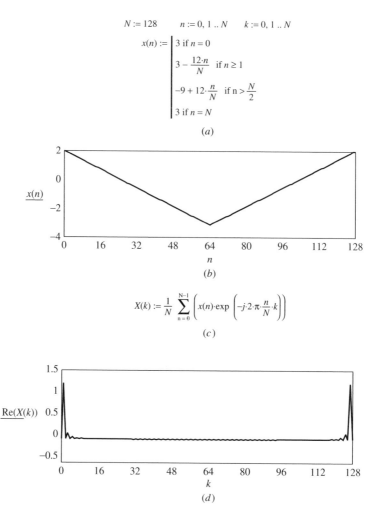

$N := 128 \qquad n := 0, 1 .. N \qquad k := 0, 1 .. N$

$$x(n) := \begin{cases} 3 \text{ if } n = 0 \\ 3 - \dfrac{12 \cdot n}{N} \text{ if } n \geq 1 \\ -9 + 12 \cdot \dfrac{n}{N} \text{ if } n > \dfrac{N}{2} \\ 3 \text{ if } n = N \end{cases}$$

(a)

(b)

$$X(k) := \frac{1}{N} \sum_{n=0}^{N-1} \left(x(n) \cdot \exp\left(-j \cdot 2 \cdot \pi \cdot \frac{n}{N} \cdot k \right) \right)$$

(c)

(d)

Figure 8-2 Hilbert transform using a triangular waveform.

the triangular wave would have. The importance of peak amplitude limits and peak power limits in circuit design must always be kept in mind.

BASIC PRINCIPLES OF THE HILBERT TRANSFORM

There are many types of transforms that are useful in electronics work. The DFT and IDFT are well known in this book because they transform back

$$xH(k) := \begin{vmatrix} -j \cdot X(k) \text{ if } k < \dfrac{N}{2} \\[2mm] 0 \text{ if } k = \dfrac{N}{2} \\[2mm] j \cdot X(k) \text{ if } k > \dfrac{N}{2} \end{vmatrix}$$

(e)

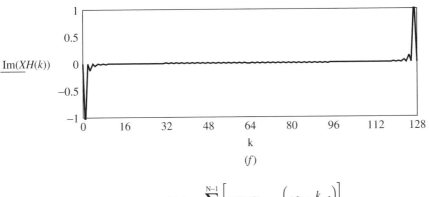

(f)

$$xh(n) := \sum_{n=0}^{N-1} \left[(XH(k) \cdot \exp\left(j \cdot 2 \cdot \pi \cdot \dfrac{k}{N} \cdot k \right) \right]$$

(g)

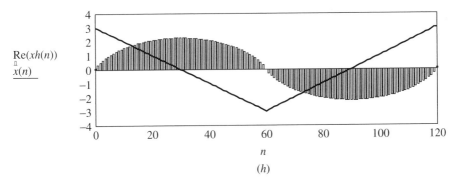

(h)

Figure 8-2 (*continued*)

and forth between the discrete time-domain signal $x(n)$ and the discrete frequency-domain spectrum $X(k)$.

Another very popular transform converts a linear differential equation into a linear algebraic equation. For example, consider the differential equation

$$v(t) = L\frac{di(t)}{dt}; \qquad \text{let } i(t) = e^{j\omega t} \text{(a phasor or a sum of phasors)}$$

$$\frac{di(t)}{dt} = j\omega e^{j\omega t} = j\omega i(t) \tag{8-1}$$

$$v(t) = Lj\omega i(t) = j\omega Li(t)$$

$$V_{ac} = j\omega L I_{ac}$$

V_{ac} and I_{ac} are sinusoidal voltage and current at frequency $\omega = 2\pi f$. The phasor $e^{j\omega t}$ is the "transformer." This is the ac circuit analysis method pioneered by Charles Proteus Steinmetz and others in the 1890s as a way to avoid having to find the steady-state solution to the linear differential equation. If the LaPlace transform is used to define a linear network (with zero initial conditions) on the S-plane, we can replace "S" with "$j\omega$, which also results in an ac circuit with sinusoidal voltages and currents. We can also start at time = zero and wait for all of the transients to disappear, leaving only the steady-state ac response. The Appendix of this book looks into this subject briefly.

These methods are today very popular and useful. If dc voltage and/or current are present, the dc and ac solutions can be superimposed.

A sum or difference of two phasors creates the cosine wave or sine wave excitation I_{ac}. These can be plugged into Eq. (8-1):

$$j\sin\omega t = \frac{e^{j\omega t} - e^{-j\omega t}}{2}, \qquad \cos\omega t = \frac{e^{j\omega t} + e^{-j\omega t}}{2} \tag{8-2}$$

The HT always starts and ends in the time domain, as shown in Figs. 8-1 and 8-2. The HT of a ($+$ sine) wave is a ($-$ cosine) wave (as in Fig. 8-1) and the ($-$ cosine) wave produces a ($-$ sine) wave. Two consecutive performances of the HT of a function followed by a polarity reversal restore the starting function.

In order to simplify the Hilbert operations we will use the phase shift method of Fig. 8-1c combined with filtering. But first we look at the basic definition to get further understanding. Consider the impulse response function $h(t) = 1/t$, which becomes infinite at $t = 0$. The HT is defined as the convolution of $h(t)$ and the signal $s(t)$ as described in Eq. (5-4) for the discrete sequences $x(m)$ and $h(m)$. The same "fold and slide" procedure is used in Eq. (8-3), where the symbol $\mathcal{H}$ means "Hilbert" and $*$ (not the same as asterisk $*$) is the convolution operator:

$$\mathcal{H}[s(t)] = \hat{s}(t) = h(t) * s(t) = \frac{1}{\pi} \int_{-\infty}^{+\infty} \frac{s(\tau)}{t - \tau} d\tau \qquad (8\text{-}3)$$

In this equation τ is the "dummy" variable of integration. The value of the integral and $\mathcal{H}[s(t)]$ become infinite when $t = \tau$ and the integral is called "improper" for this reason. First, the problem of the "exploding" integral must be corrected. This is done by separating the integral into two or more integrals that avoid $t = \tau$.

$$\begin{aligned}
\mathcal{H}[s(t)] &= \hat{s}(t) = h(t) * s(t) \\
&= \lim_{\varepsilon \to 0} \left\{ \left[\frac{1}{\pi} \int_{-\infty}^{-\varepsilon} \frac{s(\tau)}{t - \tau} d\tau + \frac{1}{\pi} \int_{+\varepsilon}^{+\infty} \frac{s(\tau)}{t - \tau} d\tau \right] \right\}
\end{aligned} \qquad (8\text{-}4)$$

This equation is called the "principal" value, also the *Cauchy principal value*, in honor of Augustin Cauchy (1789–1857). As the convolution is performed, certain points and perhaps regions must be excluded. This "connects" us with Fig. 8-1, where the value of the HT became very large at three locations.

There is also a problem if $s(t)$ has a dc component. Equations (8-3) and (8-4) can become infinite, and the dc region should be avoided. The common practice is to reduce the low-frequency response to zero at zero frequency.

The Perfect Hilbert Transformer

The procedure in Fig. 8-1c is an *all-pass network* [Van Valkenburg, 1982, Chaps. 4 and 8], also known as a *quadrature filter* [Carlson, 1986, p. 103]. Part (c) shows that its gain at all phasor frequencies, positive and negative, is ± 1.0, and that it performs an exact $+90°$ or $-90°$ phase shift. This is the practical software definition of the *perfect Hilbert transformer*.

It is useful to point out at this time that the HT of a $+$sine wave is a ($-$cosine) wave and the HT of a $+$cosine wave is a ($+$sine) wave. At a specific frequency, a $\pm 90°$ phase shift network can accomplish the same thing, but for the true HT the *wideband* constant amplitude and *wideband* constant $\pm 90°$ are much more desirable. This is a valuable improvement where these wideband properties are important, as they usually are.

In software-defined DSP equipment the almost-perfect HT is fairly easy, but in hardware some compromises can creep in. Digital integrated circuits that are quite accurate and stable are available from several vendors, for example the AD9786. In Chapter 2 we learned how to convert a two-sided phasor spectrum into a positive-sided sine$-$cosine$-\theta$ spectrum. When we are working with actual analog signal generator outputs (positive frequency), a specially designed lowpass network with an almost constant $-90°$ shift and an almost constant amplitude response over some desired *positive* frequency range is a very good component in an *analog* HT which we will describe a little later.

Please note the following: For this lowpass filter the relationship between negative frequency phase and positive-frequency phase is not simple. If the signal is a perfectly odd-symmetric sine wave (Fig. 2-2c), the positive- and negative-frequency sides are in opposite phase, just like the true HT. But if the input signal is an even-symmetric cosine wave (Fig. 2-2b) or if it contains an even-symmetric *component*, then it is not consistent with the requirements of the HT because the two sides are not exactly in opposite phase. If the signal is a random signal (or random noise), it is at least partially even-symmetric most of the time. Therefore, the lowpass filter cannot do double duty as a true HT over a two-sided frequency range, and the circuit application must work around this problem. Otherwise, the true all-pass HT is needed instead of a lowpass filter. The bottom line is that the signal-processing application (e.g., SSB) requires either an exact HT or its mathematical equivalent. Also, the validity and practical utility of the two-sided frequency concept are verified in this example.

Analytic Signal

The combination of the time sequence $x(n)$ and the time sequence $\pm j\hat{x}(n)$, where $\hat{x}(n)$, has a spectrum that occupies only one-half of the two-sided *phasor* spectrum. This is called the *analytic signal* $x\hat{a}(n)$. The result is not a physical signal that can light a light bulb [Schwartz, 1980,

p. 250]. It is a phasor spectrum that exists only in "analysis." "Analytic" also has a special mathematical meaning regarding differentiability within a certain region [Mathworld]. We have seen in Eqs. (8-3) and (8-4) that the HT does have some problems in this respect, because it is analytic only away from sudden transitions. Nevertheless, the analytic signal is a very valuable concept for us because it leads the way to some important applications, such as SSB. It is defined in Eq. (8-5), and we will soon process this "signal" into a form that is a true SSB signal that can light a light bulb and communicate.

$$xa(n) = x(n) \pm j\hat{x}(n) \tag{8-5}$$

In this equation the sequence $x(n)$ is converted to the Hilbert sequence $\hat{x}(n)$ using Eq. (8-4) shifted $\pm 90°$ by the $\pm j$ operator and added to $x(n)$.

Note that the one-sided phasor $\exp(\pm j\theta_f) = \cos\theta_f \pm j\sin\theta_f$ can be recognized as an analytic signal at any single frequency f because the HT of $\cos(\theta_f)$ is $\sin(\theta_f)$, where $\cos(\theta_f)$ and $\sin(\theta_f)$ are both real numbers. The $\pm j$ determines positive or negative frequency for this analytic signal.

Example of the Construction of an Analytic Signal

Figure 8-3 shows an example of the construction of an analytic signal. We will walk through the development.

(a) The input signal consists of two cosine waves of amplitude 1.0 and frequencies 2 and 8 (they can be any of the waves defined in Fig. 2-2).

(b) This input is plotted from $n = 0$ to $n = N - 1$ ($N = 64$). The nature of the input signal can be very difficult to determine from this "oscilloscope" display.

(c) This is the two-sided spectrum, using the DFT.

(d) The positive-frequency spectrum $X(k)$ is phase shifted $- 90°$ and the negative spectrum is shifted $+ 90°$. The $N/2$ position is set to 0. This is the Hilbert transformer.

(e) The two-sided spectrum $XH(k)$ is plotted using the DFT. The real (solid) cosine components of part (c) become imaginary (dotted) sine components in part (e).

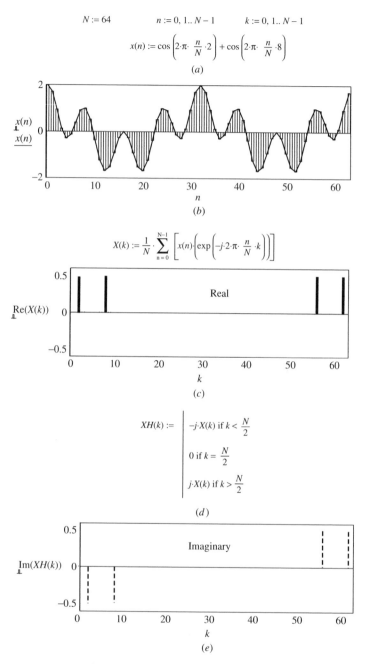

$$N := 64 \qquad n := 0, 1..N-1 \qquad k := 0, 1..N-1$$

$$x(n) := \cos\left(2 \cdot \pi \cdot \frac{n}{N} \cdot 2\right) + \cos\left(2 \cdot \pi \cdot \frac{n}{N} \cdot 8\right)$$

(a)

(b)

$$X(k) := \frac{1}{N} \cdot \sum_{n=0}^{N-1} \left[x(n) \cdot \left(\exp\left(-j \cdot 2 \cdot \pi \cdot \frac{n}{N} \cdot k\right) \right) \right]$$

(c)

$$XH(k) := \begin{vmatrix} -j \cdot X(k) \text{ if } k < \frac{N}{2} \\[2mm] 0 \text{ if } k = \frac{N}{2} \\[2mm] j \cdot X(k) \text{ if } k > \frac{N}{2} \end{vmatrix}$$

(d)

(e)

Figure 8-3 The analytic signal.

$$xh(n) := \sum_{k=0}^{N-1}\left[XH(n)\cdot\exp\!\left(-j\cdot 2\cdot\pi\cdot\frac{k}{N}\cdot n\right)\right]$$

(f)

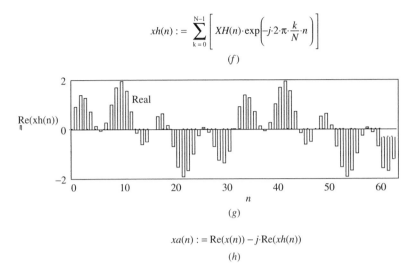

(g)

$$xa(n) := \operatorname{Re}(x(n)) - j\cdot\operatorname{Re}(xh(n))$$

(h)

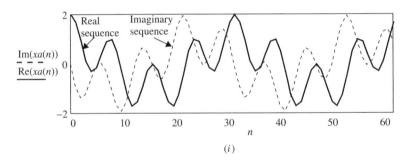

(i)

$$XA(k) := \frac{1}{N}\cdot\sum_{n=0}^{N-1}\left(xa(n)\cdot\exp\!\left(-j\cdot 2\cdot\pi\cdot\frac{n}{N}\cdot k\right)\right)$$

(j)

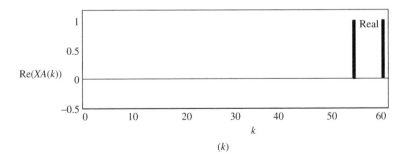

(k)

Figure 8-3   *(continued)*

(f) The spectrum $XH(k)$ is converted to the time domain $xh(n)$ using the IDFT. This is the Hilbert transform HT of the input signal $x(n)$.

(g) The HT $xh(n)$ of the input signal sequence is plotted. Note that $xh(n)$ is a real sequence, as is $x(n)$.

(h) The formula $xa(n)$ for the complex analytic signal in the time domain.

(i) There are two time-domain plot *sequences*, one dashed for the imaginary part of $xa(n)$ and one solid for the real part of $xa(n)$. These I and Q *sequences* are in phase quadrature.

(j) The spectrum $XA(k)$ of the analytic signal is calculated.

(k) The spectrum of the analytic signal is plotted. Only the negative-frequency real components -2 (same as 62) and -8 (same as 56) appear because the minus sine was used in part (H). If the plus sign were used in part (h), only the positive-frequency real components at $+2$ and $+8$ would appear in part (k). Note that the amplitudes of the frequency components are twice those of the original spectrum in part (c). All of this behavior can be understood by comparing parts (c) and (e), where the components at 2 and 8 cancel and those at -2 and -8 add, but *only* after the equation in part (h) is used. The $\pm j$ operator in part (h) aligns the components in the correct phase either to augment or to cancel.

This is the baseband analytic signal, also known as the *lowpass equivalent spectrum* [Carlson, 1986, pp. 198–199] that is centered at zero frequency. To use this signal, for example in radio communication, it must be frequency-translated. It then becomes a true single-sideband "signal" at positive SSB frequencies with suppressed carrier. If this SSB RF signal is represented as phasors, it is a two-sided SSB phasor, spectrum, one SSB sideband at positive RF frequencies and the other SSB sideband at negative RF frequencies. The value of the positive suppressed carrier frequency ω_0 can be anything, but in the limit, as $\omega_0 \to 0$, the idea of an actual SSB signal disappears (in principle), as mentioned before.

SINGLE-SIDEBAND RF SIGNAL

At radio frequencies the single-sideband (SSB) signal contains information only on upper singlesideband (USSB) or only on lower singlesideband

(LSSB). The usual "carrier" that we see in conventional AM is missing. A related approach is the "vestigial" opposite sideband, which tapers off in a special manner. Some systems (e.g., shortwave broadcast) use a reduced-level *pilot carrier* ($-12\,\text{dB}$), that is used to phase lock to an input signal. In the usual peak-power-limited system the power in the pilot carrier reduces slightly the power in the desired single sideband. Special methods are used to modulate and demodulate the SSB signal, which add somewhat to the cost and complexity of the equipment. The big plus factor is that almost all of the transmitted and received signal power reside in one narrow sideband, where they are most effective.

Incidentally, and to digress for a moment, AM mode is criticized because of all of the "wasted" power that is put into the carrier. Nothing could be further from the truth. The basic AM receiver uses a very simple diode detector for the AM signal. The AM carrier is the "local oscillator" for this detector which demodulates (translates) the AM signal to audio (see Fig. 3-5). The transmitted carrier can service many millions of simple AM receivers in this manner. The receiver provides the power level at the carrier frequency that the detector requires and provides synchronous demodulation. In advanced designs a -12-dBc PLL pilot carrier, possibly combined with SSB, is created that reduces "selective fading" (search "selective fading" and subtopics such as "OFDM" on the Web).

We will use the baseband analytic signal $xb(n)$ to create mathematically in Fig. 8-4 a high-frequency SSB two-tone USSB or LSSB power spectrum that is capable of radio communication. Figure 8-4g is the desired USSB output.

(a) $klo = 10$ is the frequency of the carrier that is to be suppressed in SSB.

(b) Part (b) is the time sequence $x(n)$ of the two-tone baseband inputs at frequencies 9 and 12.

(c) Part (c) is the two-sided baseband spectrum $X(k)$ of the two-tone baseband input.

(d) Part (d) is the ideal Hilbert transform $XH(k)$ of the two-tone baseband input.

(e) The analytic spectrum $[X(k) + jXH(k)]$ is formed and then multiplied by the carrier frequency $klo = 10$. This frequency-converted result is

$$N := 64 \qquad n := 0, 1..N-1 \qquad k := 0, 1,..N-1 \qquad klo := 10$$

(a)

$$x(n) := \cos\left(2 \cdot \pi \cdot \frac{n}{N} 9\right) + \sin\left(2 \cdot \pi \cdot \frac{n}{N} \cdot 12\right)$$

(b)

$$X(k) := \frac{1}{N} \cdot \sum_{n=0}^{N-1} \left(x(n) \cdot \exp\left(-j \cdot 2 \cdot \pi \cdot \frac{n}{N} \cdot k\right)\right)$$

(c)

$$XH(k) := \begin{vmatrix} -j \cdot X(k) \text{ if } k < \frac{N}{2} \\[2mm] 0 \text{ if } k = \frac{N}{2} \\[2mm] j \cdot X(k) \text{ if } k > \frac{N}{2} \end{vmatrix}$$

(d)

$$xb(n) := \sum_{n=0}^{N-1} \left[[X(k) + j \cdot XH(k)] \cdot \exp\left(j \cdot 2 \cdot \pi \cdot \frac{n}{N} \cdot klo\right) \cdot \exp\left(j \cdot 2 \cdot \pi \cdot \frac{n}{N} \cdot k\right) \right]$$

(e)

$$XB(k) := \frac{1}{N} \cdot \sum_{n=0}^{N-1} \left[xb(n) \cdot \exp\left(-j \cdot 2 \cdot \pi \cdot \frac{n}{N} \cdot k\right) \right]$$

(f)

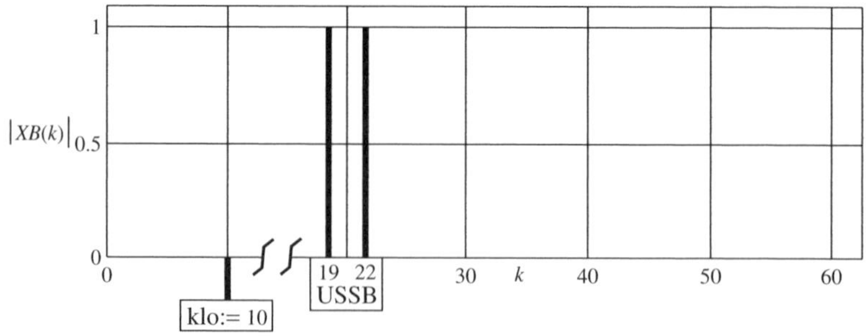

(g)

Figure 8-4 Construction of an upper single-sideband signal.

then converted to the time-domain signal $xb(n)$, centered at $k = 10$, using the IDFT.

(f) The DFT of $xb(n)$ provides the two-tone *real* signal at $k = 19$ and 22. This is the spectrum $XB(k)$ of the USSB. RF frequencies are 19 and 22. The LSSB output is obtained by using $[X(k)-jXH(k)]$ in step (e).

(g) Part (g) is the frequency plot of the USSB RF signal. Note the absence of any outputs except the desired two-tone *real* signal. Only the magnitudes of the two outputs are of interest in this experiment, but the phase of each can also be plotted.

Note that, as always, the multiplication of two time-domain sequences [part (f)] is a nonlinear process, and the subsequent DFT then reveals a spectrum of two real signals. The SSB output is a real signal, not an analytic signal as defined in Eq. (8-5).

Figure 8-4 is not a realistic example in terms of actual baseband and RF frequencies, but the general idea is conveyed correctly. Figures 8-1 and 8-2 may also be reviewed as needed. The reader can use the time- and frequency-scaling procedures in Chapter 1 and some appropriate graph method (if needed) to get the real-world model working.

SSB DESIGN

It is interesting to look briefly at some receiving and transmitting methods that can be implemented using analog or discrete-signal methods. Comparable methods are used in DSP [Sabin and Schoenike, 1998, Chap. 8].

The basic op-amp first-order all-pass network shown in Fig. 8-5 has a constant gain *magnitude* ≈ 1.0 at all frequencies, including zero, phase $\approx +180°$ at very low frequency, and phase $\approx 0°$ at very high frequency. The output phase is $90°$ *leading* at $\omega = a$ (confirmed by analysis and simulation). The gain at zero frequency is -1.0, which corresponds to $+180°$. At zero frequency on the S-plane, the pole on the left-side for this "nonminimum phase" network is at $0°$ and the zero on the right side

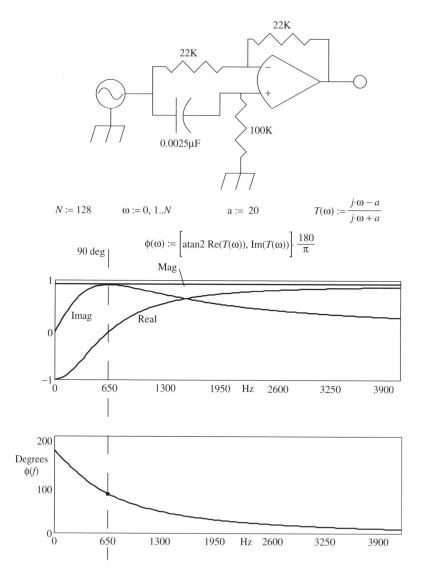

$N := 128$ $\quad$ $\omega := 0, 1..N$ $\quad\quad$ $a := 20$ $\quad\quad$ $T(\omega) := \dfrac{j \cdot \omega - a}{j \cdot \omega + a}$

$$\phi(\omega) := \left[\mathrm{atan2}\ \mathrm{Re}(T(\omega)),\ \mathrm{Im}(T(\omega)) \right] \cdot \frac{180}{\pi}$$

Figure 8-5 Elementary all-pass active *RC* network.

is at $+180°$, according to the usual conventions [Dorf, 1990, Figs. 7-15b and 7-16].

$$T(j\omega) = \frac{j\omega - a}{j\omega + a}, \ T(s) = \frac{s - z}{s + p}$$

Note the use of the Mathcad function atan2(x, y) that measures phase out to $\pm 180°$ (see also Chapter 2). The values 0.0025 μF and 100 K are modified in each usage of this circuit. Metal film resistors and stable NP0 capacitors are used. The op-amp is of high quality because several of them in cascade are usually dc coupled.

Figure 8-6 shows how these basic networks can be combined to produce a wideband $-90°$ phase shift with small phase error and almost constant amplitude over a baseband frequency range. Each of the two all-pass networks (I and Q) is derived from a computer program that minimizes the phase *error* between the I and Q channels on two separate "wires." [Bedrosian, 1963] is the original and definitive IRE article on this subject. Examples of the circuit design and component values of RC op-amp networks are in [Williams and Taylor, 1995, Chap. 7] and numerous articles. A simulation of this circuit from 300 to 3000 Hz using Multisim and the values from the book of Williams and Taylor (p. 7.36) shows a maximum phase error of 0.4°. The 6 capacitors are 1000 pF within 1.0%. The input and output of each channel may require voltage-follower op-amps to assure minimal external loading by adjacent circuitry. Copying R and C values from a handbook in this manner is sometimes quite sensible when the alternatives can be unreasonably labor-intensive. A high-speed PC could possibly be used to fine-tune the phase error in a particular application (see, for example, [Cuthbert, 1987], and also Mathcad's optimizing algorithms).

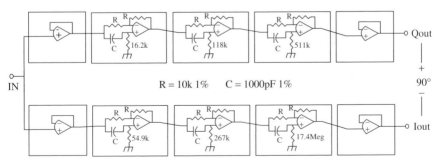

Figure 8-6 Two sets of basic all-pass networks create I and Q outputs with a 90° phase difference across the frequency range 300 to 3000 Hz.

The following brief discussion provides some examples regarding the usage of the Hilbert transform and its mathematical equivalent in radio equipment. Analog methods are used for visual convenience.

SSB TRANSMITTER

We illustrate in Fig. 8-7 the analog design of an SSB transmitter signal using the phase-shift method. It uses the $-90°$ lowpass (positive-frequency) filter of Fig. 8-6, two double-balanced mixers, and an HF local oscillator [Krauss et al., 1980, Chap. 8]. The mixers create two double-sideband suppressed carrier (DSBSC) signals. The combiner at the output uses the sum of these two inputs to create at the local oscillator frequency ω_0 an LSB or the difference of the two inputs to create an USB. The BPF restricts the output to some desired frequency band. The end result is equivalent mathematically to a synthesis of the Hilbert transform and the analytic signal translated to RF that we have considered in this chapter.

There is an interesting artifact of this circuit that we should look at.

1. Start at the input, where the baseband signal is $\cos \omega_m t$ at $0°$ reference.

2. The I-channel output (a) has a phase shift $\angle\theta°$, relative to the $0°$ reference input, that varies from $+64°$ at 300 Hz to $-154°$ at 3 kHz. The I-channel output (a) is $\cos\omega_m t + \theta°$. This effect is inherent in the design of this filter.

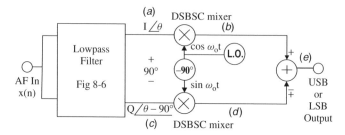

Figure 8-7   SSB generator using the phasing method.

3. Because the wideband phase shift from 300 to 3000 Hz is very nearly $-90°$ from I to Q, the Q output (c) has the same additional shift $\theta°$ as the I-channel output (a).

If we compare locations (a) and (c) we see that they differ only in phase and not in frequency. So this process is not *phase modulation*, which would have to be a nonlinear process that creates phase modulation sidebands. It is an additive process that does not contribute additional spectrum components. For a typical SSB speech signal this phase shift is usually not noticed by a human listener, although some amplitude modification (not the same as nonlinear distortion) can occur if the circuitry is not almost linear-phase. It could be noticed in data modes that are not normally used in SSB. The important thing is that the I and Q channels are separated by very nearly $90°$, positive at the I channel and negative at the Q channel.

In a DSP SSB transmitter an FIR design HT would need only a single channel, located, for example, on the Q side [Sabin and Schoenike, 1998, Chap. 8].

Also, other phase errors in the circuit can reduce the degree of cancellation of the undesired sideband. A practical goal for this cancellation is in the range 40 to 50 dB.

FILTER METHOD TRANSMITTER

Figure 8-8 shows the filter method of creating an SSB signal. The DSBSC signal goes through a narrowband mechanical or crystal filter. The filter creates the one-sided real SSB signal at IF, and the result is indistinguishable from the phasing method. Both methods are basically equivalent mathematically in terms of the analytic signal [Carlson, 1986, Chap. 6]. In other words, the result of a frequency translation of the transmit signal to baseband is indistinguishable from the analytic signal in Eq. (8-5.)

PHASING METHOD SSB RECEIVER

Figure 8-9 illustrates a phase-shift, image-canceling SSB receiver. It is similar to the SSB transmitter except that two identical lowpass filters are

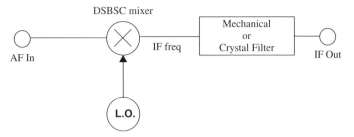

Figure 8-8 SSB generator using the IF filter method.

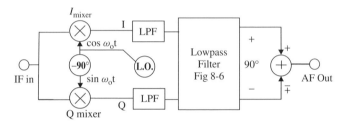

Figure 8-9 Phasing method image-canceling SSB demodulator.

used after the IF or RF down-conversion to baseband (especially in the direct-conversion receiver) to establish the desired audio-frequency range and attenuate undesired mixer outputs that can interfere with the desired input frequency range. The lowpass filter of Fig. 8-6 provides the I and Q audio. The combiner selects the USB or LSB mode. The mixers are identical double-balanced types that perform the DSBSC function. Digital circuitry that divides four times the desired L.O. frequency by four and also provides two quadrature outputs, I_{LO} and Q_{LO}, is frequently used [Sabin and Schoenike, 1998, Chap. 4], especially when the L.O. frequency must be variable to cover an input signal range.

FILTER METHOD RECEIVER

Figure 8-8, flipped from left to right, shows the receiver IF filter method. The narrowband filter precedes the down-converter mixer. This method is also equivalent to the phasing method, which has a possible advantage in circuit cost, where crystal and mechanical filters are usually more expensive. Receivers very often combine the phasing and filter methods in the same or different signal frequency ranges to get greatly improved performance in difficult-signal environments.

The comments for the SSB transmitter section also apply to the receiver, and no additional comments are needed for this chapter, which is intended only to show the Hilbert transform and its mathematical equivalent in a few specific applications. Further and more complete information is available from a wide variety of sources [e.g., Sabin and Schoenike, 1998], that cannot be pursued adequately in this introductory book, which has emphasized the analysis and design of *discrete* signals in the time and frequency domains.

REFERENCES

Bedrosian, S. D., 1963, Normalized design of $90°$ phase-difference networks, *IRE Trans. Circuit Theory*, vol. CT-7, June.

Carlson, A. B., 1986, *Communication Systems*, 3rd ed., McGraw-Hill, New York.

Cuthbert, T. R., 1987, *Optimization Using Personal Computers with Applications to Electrical Networks*, Wiley-Interscience, New York. See trcpep@aol.com or used-book stores.

Dorf, R. C., 1990, *Modern Control Systems*, 5th ed., Addison-Wesley, Reading, MA, p. 282.

Krauss H. L., C. W. Bostian, and F. H. Raab, 1980, *Solid State Radio Engineering*, Wiley, New York.

Mathworld, http://mathworld.wolfram.com/AnalyticFunction.html.

Sabin, W. E., and E. O. Schoenike, 1998, *HF Radio Systems and Circuits*, SciTech, Mendham, NJ.

Schwartz, M., 1980, *Information Transmission, Modulation and Noise*, 3rd ed., McGraw-Hill, New York.

Van Valkenburg, M. E., 1982, *Analog Filter Design*, Oxford University Press, New York.

Williams, A. B., and F. J. Taylor, 1995, *Electronic Filter Design Handbook*, 3rd ed., McGraw-Hill, New York.

APPENDIX

Additional Discrete-Signal Analysis and Design Information[†]

This brief Appendix will provide a few additional examples of how Mathcad can be used in discrete math problem solving. The online sources and Mathcad *User Guide* and Help (F1) are very valuable sources of information on specific questions that the user might encounter in engineering and other technical activities. The following material is guided by, and is similar to, that of Dorf and Bishop [2004, Chap. 3].

DISCRETE DERIVATIVE

We consider first Fig. A-1, the *discrete derivative*, which can be a useful tool in solving discrete differential equations, both linear and nonlinear. We consider a specific example, the exponential function exp($\cdot$) from

---

[†]Permission has been granted by Pearson Education, Inc., Upper Saddle River, NJ, to use in this appendix, text and graphical material similar to that in Chapter 3 of [Dorf and Bishop, 2004].

Discrete-Signal Analysis and Design, By William E. Sabin
Copyright © 2008 John Wiley & Sons, Inc.

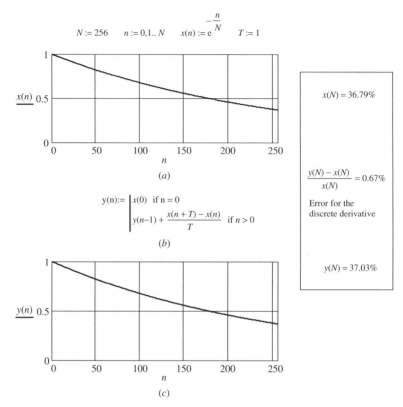

Figure A-1 Discrete derivative: (a) exact exponential decay; (b) definition of the discrete derivative; (c) exponential decay using the discrete derivative.

$n = 0$ to $N - 1$ that decays as

$$x(n) = \exp\left(\frac{-n}{N}\right), \quad 0 < n < N - 1 \qquad \text{(A-1)}$$

The decay of this function from $n = 0$ to N is from 1.0 to 0.3679, corresponding to a time constant of 1.0. Figure A-1 shows the exact decay.

Now consider the discrete approximation to this derivative, called $y(n)$, and define $\Delta y(n)/\Delta n$ as an *approximation* to the true derivative, as follows:

$$y(n) = \begin{cases} x(0) & \text{if } n = 0 \\ y(n-1) + \frac{x(n+T)-x(n)}{T} & \text{if } n > 0 \end{cases} \qquad (A\text{-}2)$$

$T = 1$ in this example.

In this equation the second additive term is derived from an increment of $x(n)$. In other words, at each step in this process, $y(n)$ hopefully does not change too much (in some situations with large sudden transitions, it might). The advantage that we get is an easy-to-calculate discrete approximation to the exact derivative.

Figure A-1c shows the decay of $x(n)$ using the discrete derivative. In part (b) the accumulated error in the approximation is about 0.67%, which is pretty good. Smaller values of T can improve the accuracy; for example, $T = 0.1$ gives an improvement to about 0.37%, but values of T smaller than this are not helpful for this example. A larger number of samples, such as 2^9, is also helpful. The discrete derivative can be very useful in discrete signal analysis and design.

STATE-VARIABLE SOLUTIONS

We will use the discrete derivative and matrix algebra to solve the two-state differential equation for the *LCR* network in Fig. A-2. There are two energy storage elements, L and C, in the circuit. There is a voltage across and a displacement current through the capacitor C, and a voltage across and an electronic current through the inductor L. We want all of these as a function of time t. There are also possible initial conditions at $t = 0$, which are a voltage V_{C0} on the capacitor and a current I_{L0} through the inductor, and a generator (u) (in this case, a current source) is connected as shown. The two basic differential equations are, in terms of v_C and i_L,

$$N := 8 \qquad n := 0,1..N \qquad T := 0.1$$

$$V_C := 1.5 \qquad IL := 1.0 \qquad R := 3 \qquad L := 1 \qquad C := 0.5$$

$$x(n) := \begin{vmatrix} \begin{pmatrix} V_C \\ IL \end{pmatrix} & \text{if } n = 0 \\[2em] \left[T \cdot \begin{pmatrix} 0 & \dfrac{-1}{C} \\[1em] \dfrac{1}{L} & \dfrac{R}{L} \end{pmatrix} + \begin{pmatrix} 1 & 0 \\ 0 & 1 \end{pmatrix} \right] x(n-1) & \text{if } n > 0 \end{vmatrix}$$

Solution to matrix differential equation for initial conditions of $V_C = 1.5$, $I_L = 1.0$

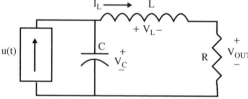

Figure A-2   *LCR* Circuit differential equation solution for initial values of V_C and I_L, $I_{gen} = 0$.

$$i_C = C\frac{dv_C}{dt} = u - i_L$$

$$v_L = L\frac{di_L}{dt} = v_C - Ri_L \qquad \text{(A-3)}$$

$$v_{OUT} = Ri_L$$

Rewrite Eq. (A-3) in state-variable format:

$$\dot{v_C} = 0v_C - \frac{1}{C}i_L + \frac{1}{C}u$$

$$\dot{i_L} = \frac{1}{L}v_C - \frac{R}{L}i_L + 0u \qquad \text{(A-4)}$$

$$v_O = Ri_L$$

A nodal circuit analysis confirms these facts for this example. R, L, and C are constant values, but they can easily be time-varying and/or nonlinear functions of voltage and current. The discrete analysis method deals with all of this very nicely.

We now add in the initial conditions at time zero, V_{C0} and I_{L0}:

$$\dot{v_C} = 0(v_C + V_{C0}) - \frac{1}{C}(i_L + I_{L0}) + \frac{1}{C}u$$

$$\dot{i_L} = \frac{1}{L}(v_C + V_{C0}) - \frac{R}{L}(i_L + I_{L0}) + 0u \qquad \text{(A-5)}$$

$$v_O = Ri_L$$

The two derivatives appear on the left side. Note that if $(v_C + V_{C0})$ is multiplied by zero, the rate of change of v_C does not depend on that term, and the rate of change of i_L does not depend on u if the u is multiplied by zero. The options of Eqs. (A-4) and (A-5) can easily be imagined. Description of flow-graph methods in [Dorf and Bishop, 2004, Chaps. 2 and 3] and in numerous other references are excellent tools that are commonly used for these problems. We will not be able to get deeply into that subject in this book, but Fig. A-4 is an example.

The next step is to rewrite Eq. (A-5) in matrix format. Also, v_C is now called X_1, and I_L is now called X_2.

$$\begin{pmatrix} \dot{X_1} \\ \dot{X_2} \end{pmatrix} = \begin{pmatrix} 0 & \frac{-1}{C} \\ \frac{1}{L} & \frac{-R}{L} \end{pmatrix} \begin{pmatrix} X_1 \\ X_2 \end{pmatrix} + \begin{pmatrix} \frac{1}{C} \\ 0 \end{pmatrix} (u) \qquad \text{(A-6)}$$

$$V_O = RX_2$$

Now write the (A-6) equations as follows:

$$\dot{X}_1 = 0X_1 - \frac{1}{C}X_2 + \frac{1}{C}u$$

$$\dot{X}_2 = \frac{1}{L}X_1 - \frac{R}{L}X_2 + 0u \tag{A-7}$$

$$V_O = RX_2$$

Next, we will solve Eq. (A-6) [same as Eq. (A-7) for X_1 ($=v_c$), X_2 ($=i_L$), and V_O]. Component u is the input signal generator.

This general idea applies to a wide variety of practical problems (see [Dorf and Bishop, 2004, Chap. 3] and many other references). The methods of matrix algebra and matrix calculus operations are found in many handbooks (e.g., [Zwillinger, 1996]).

The general format for state-variable equations, similar to Eqs. (A-4) and (A-5), is

$$\begin{aligned}\dot{\mathbf{x}} &= \mathbf{Ax} + \mathbf{Bu} \\ \mathbf{y} &= \mathbf{Cx} + \mathbf{Du}\end{aligned} \tag{A-8}$$

in which **A, B, C**, and **D** are coefficient matrices whose numbers may be complex and varying in some manner with time, voltage, or current, **u** pertains to complex signal sources, and **y** is the complex output for a complex input signal **u**. The boldface letters denote matrices specifically. Comparing Eq. (A-6) with Eq. (A-5), the general idea is clear. At any time t, **x** is the collection (vector) of voltages v and currents i in the circuit, and $\dot{\mathbf{x}}$ is the collection (vector) of their time derivatives.

Matrix algebra or matrix calculus, using Mathcad, finds all the values of x and y at each value of time t with great rapidity, and plots graphs of the results. Starting at $t = 0$, from a set of initial values of V and I, we can trace the history of the network through the transient period and into the steady state.

We will look more closely at the general method of matrix construction for our specific example. Equations (A-6), (A-7), and (A-9) can be Laplace-transformed. In this process the unit matrix $[I] = \begin{bmatrix} 1 & 0 \\ 0 & 1 \end{bmatrix}$, and this work must be in accordance with the rules of matrix algebra and the

Laplace transform:

1. $sX(s) - x(0) = AX(s) + BU(s)$

2. $sX(s) - AX(s) = x(0) + BU(s)$

3. $X(s)[sI - A] = x(0) + BU(s)$

4. $X(s) = [sI - A]^{-1}x(0) + [sI - A]^{-1}BU(s)$

(A-9)

This can be inverse-Laplace-transformed to get $x(t)$, a function of time for each $X(s)$ [Dorf and Bishop, 2004, Sec. 3], but we want to use the discrete derivative of Eq. (A-2) as an alternative for discrete-signal analysis and design [Dorf and Bishop, 2004, Sec. 3].

USING THE DISCRETE DERIVATIVE

We now replace the $\dot{x}$ matrix in Eq. (A-6) by incorporating Eq. (A-2). Using the intermediate steps in Eq. (A-10) and using the sequence index $x(n)$ method that we are already very familiar with produces the following Mathcad program, expressed in Word for Windows format.

1. (if $n = 0$) $\begin{pmatrix} \mathbf{x_0(n)} \\ \mathbf{x_1(n)} \end{pmatrix} = \begin{pmatrix} V_{C0} \\ I_{L0} \end{pmatrix}$

2. (if $n > 0$) $\begin{pmatrix} \mathbf{x_0(n)} \\ \mathbf{x_1(n)} \end{pmatrix} = \left[T \begin{pmatrix} 0 & \frac{-1}{C} \\ \frac{1}{L} & \frac{-R}{L} \end{pmatrix} + \begin{pmatrix} 1 & 0 \\ 0 & 1 \end{pmatrix} \right]$ (A-10)

$\times \begin{pmatrix} \mathbf{x_0(n-1)} \\ \mathbf{x_1(n-1)} \end{pmatrix} + T \begin{pmatrix} b_0 \\ b_1 \end{pmatrix} \mathbf{u(n)}$

Figure A-3 shows this equation in Mathcad program form. We can immediately use three options:

1. Initial conditions $x_0(0)$ and $x_1(0)$ can be zero, $u(n)$ can have a dc value or a time function such as a step or sine wave, and in this example [Eqs. (A-6) and (A-7)] $b_0 = 1$, $b_1 = 0$, $T = 0.1$, $u(n) = 1.0$.
2. $u(n)$ can be zero, and the transient response is driven by $x(0)$, an initial value of capacitor voltage or inductor current or both.

$N := 1000$ $\qquad$ $n := 0,1.. N$ $\qquad$ $T := 0.01$ $\qquad$ $R := 0.01$ $\qquad$ $L := 0.5$ $\qquad$ $C := 0.5$

$$b := \begin{pmatrix} 1 \\ 0 \end{pmatrix} \qquad\qquad u(n) := 100 \sin\left(2\,\pi\,\frac{n}{N}\,4\right) mA$$

$$x(n) := \left| \begin{array}{l} \begin{pmatrix} 0 \\ 0 \\ 0 \end{pmatrix} \text{ if } n = 0 \\[2em] \left[T\left(\begin{pmatrix} 0 & \dfrac{-1}{C} \\ \dfrac{1}{L} & \dfrac{R}{L} \end{pmatrix} + \begin{pmatrix} 1 & 0 \\ 0 & 1 \end{pmatrix} \right) . x(n-1) + T\,b\,u(n) \text{ if } n > 0 \right] \end{array} \right.$$

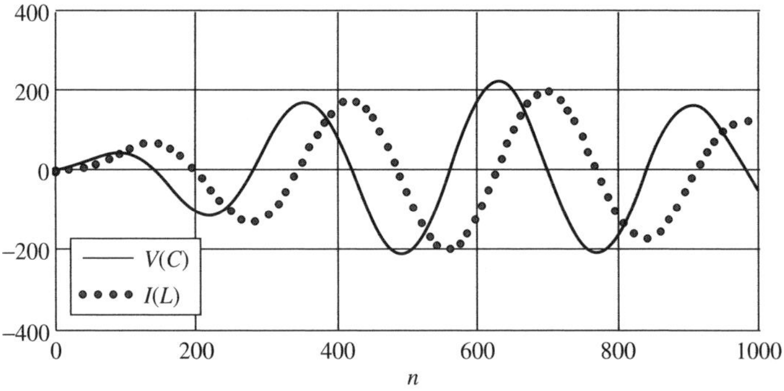

Figure A-3   Time response of the *LCR* network with zero initial conditions and sine-wave current excitation.

3. $x(0)$ and $u(n)$ can both be operational at $t = 0$ or later than zero. There are a lot of options for this problem.

The Mathcad worksheet Fig. A-2 shows option 2 for initial condition values of V_C and I_L, and $u(n) = 0$. The values of $x(n)_0$ and $x(n)_1$ are plotted.

Figure A-3 uses zero initial conditions and $u(n) = 100\,mA$ sine wave, frequency $= 4$, $b_0 = 1.0$, $b_1 = 0$. The buildups from zero of x_1 (capacitor volts) and x_2 (inductor amperes) are plotted. The lag of I_L (peaks at a later time) can be noticed.

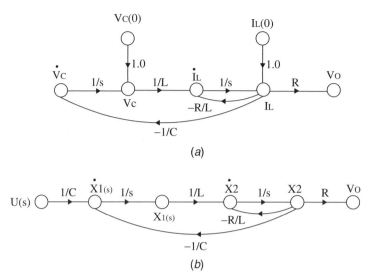

Figure A-4   Flow-chart for the network of Fig. A-2: (a) with no input (u) but with initial values of V_C and I_L; (b) with no initial conditions but with a sine-wave input signal $u(t)$.

The book by [Dorf and Bishop] explores this problem using several different methods that are very instructional but that we do not pursue in this book. The reader is encouraged to become more familiar with the network analysis methods described in this appendix. It is good practical engineering.

Finally, Fig. A-4 illustrates the two varieties of flow graph for the network discussed in this appendix. We can understand Fig. A-4a by referring to Eq. (A-5) with u set to zero (no external inputs) and with initial values of $V_C(0)$ and $I_L(0)$, as shown also in Fig. A-2. In Fig. A-4b, V_C, I_L, and their derivatives correspond to those in Eq. (A-5) with initial conditions V_C and I_L set to zero, as shown in Fig. A-3, and the input u drives the network from a zero start with a sine wave that starts at zero value. The output peak amplitude $V_O(t)$ fluctuates for at least the 1000 time increments illustrated.

It is also an interesting exercise for the reader to calculate and plot the inductor voltage and current and the capacitor voltage and current as functions of time n in Figs. A-2 and A-3.

REFERENCES

Dorf, R. C., and R. H. Bishop, 2004, *Modern Control Systems*, 10th ed., Prentice Hall, Upper Saddle River, NJ, Chap. 3.

Zwillinger, D., Ed., 1996, *CRC Standard Mathematical Tables and Formulae*, 30th ed., CRC Press, Boca Raton, FL.

GLOSSARY

Adjacent channel interference. One or more adjacent channel signals create interference in a desired channel by aliasing or wideband emissions.

Aliasing (classical). In positive-only frequency systems, a signal in part of the positive-frequency region is invaded by a second signal that is in an adjacent part of the positive-frequency region.

Aliasing. The overlapping (invasion) from one 0 to $N-1$ time or frequency sequence to an adjacent 0 to $N-1$ time or frequency sequence.

Amplitude noise. Noise created by variations in the amplitude of a signal.

Analytic signal (sequence). An $X(k)$, its Hilbert transform $\hat{X}(k)$ and the $\pm j$ operator combine to create a phasor sequence that is one-sided in the positive- or negative-frequency domain. The phasor $A \exp(\pm j\theta)$ is an analytic signal. The analytic phasor sequence is used to construct SSB signals digitally or discretely. It is synthesized to design analog SSB systems.

Auto-covariance. The ac component of an autocorrelation.

Average value. The time average of a signal between two time limits, often minus infinity to plus infinity.

Discrete-Signal Analysis and Design, By William E. Sabin
Copyright © 2008 John Wiley & Sons, Inc.

Boltzmann's constant. 1.38×10^{-23} joules per Kelvin. Used in noise calculations.

Coherent. Two time signals $x_1(n)$ and $x_2(n)$ are coherent if their $x(n)$ values add together algebraically at each (n). In the frequency domain the $X(k)$s add in a similar manner.

Complex frequency domain. Values of $X(k)$ phasors contain a real part, an imaginary part, an amplitude value, a frequency value, and a phase value relative to some reference phase value. The domain has a positive-frequency side and an equal-length negative-frequency side.

Complex plane. The two-dimensional rectangular plane of the real axis (x) and the imaginary axis (jy) (see Fig. 1-5).

Complex signal. A signal that is defined as part real and part imaginary on the complex plane. In the time domain, *sequences* can be complex. In the frequency domain, a single phasor can be complex.

Convolution. A fold, slide, and multiply operation to obtain an overlap area between two geometric or mathematical regions.

Correlation. A measure of the similarity of a function and a time- or frequency-shifted copy of the function (auto correlation) or the similarity of two different functions, one of which is shifted (cross-correlation).

Correlation coefficient. A measure of the "relatedness" in some sense, from -1 to $+1$, of two nondeterministic or deterministic processes.

Cross-covariance. The ac component of a cross-correlation.

Cross power spectrum. The commonality of power spectrum in two associated signals.

Discrete derivative. An approximate implementation of a time-derivative that uses the discrete sequence $x(n)$.

Discrete Fourier series. In discrete-signal length-N analysis, a periodic repeating waveform can be defined as a useful set of positive-frequency harmonics from $k = 1$ to $k = N/2 - 1$.

Discrete Fourier transform (DFT). Converts the time domain $x(n)$ to the frequency domain $X(k)$.

Discrete Fourier transform of convolution. Converts a convolution of two time sequences to the product of two frequency sequences: the system function. Used in linear system analysis.

Discrete frequency. Signals $X(k)$ in the frequency domain occur at discrete values of frequency (k) from 0 to $N - 1$.

Discrete time. Signals $x(n)$ in the time domain occur at discrete values of time (n) from 0 to $N - 1$.

Digital signal processing (DSP). Signal processing in which signal amplitudes are also discrete (quantized).

Even symmetry. The two sides, $X(k)$ and $X(N - k)$, of a phasor spectrum have the same phase.

Expected value. The sum of products of a signal amplitude at time T and the probability of occurrence of the signal at time T [Eq. (6-1)]. Also known as the *first moment*.

Fast Fourier transform (FFT). A high-speed algorithm for the DFT.

Flow graph. A graphical method of tracing the flow of signals within a network.

Fourier, Joseph. French mathematician who originated the trigonometric series method of analysis and design of mathematical and physical phenomena.

Frequency domain. Signals are classified according to their occurrence in frequency (f) continuous or discrete $X(k)$.

Frequency scaling. A sequence of frequency values have a certain sequential relationship from low end to high end. The maximum frequency minus the minimum frequency, divided by the number of frequencies, is the frequency scale factor.

Gaussian noise. Random electrical noise, perhaps thermally generated noise, that has the Gaussian (normal) amplitude probability density function.

Hermitian symmetry. A spectral property such that positive- and negative-frequency values are complex conjugates. The sine and cosine-wave phasors are Hermitian

Hilbert transform. In RF work, an algorithm that modifies a two-sided phasor spectrum so that positive-frequency phasors are phase shifted $-90°$ and negative-frequency phasors are phase shifted $+90°$. This idea is useful in many applications, especially in SSB.

Integer. A collection of whole numbers: such as $\pm(1, 2, 3, \ldots)$.

Intermodulation. Two or more input signals combine in a nonlinear circuit or device to create spurious output frequencies.

Inverse discrete Fourier transform (IDFT). Converts the frequency domain $X(k)$ to the time domain $x(n)$. Defined according to Bracewell.

Inverse fast Fourier transform (IFFT). A high-speed alternative for the IDFT. A Mathcad function defined according to Bracewell.

Laplace transform. Converts a function in the S-plane $\sigma \pm j\omega$ domain to a function in the time domain. The inverse transform performs the opposite process.

Mathcad. A personal computer program that performs a very wide range of mathematical calculations, either numerical or symbolic, in interactive form.

Mathcad program. A structured set of logical operations that perform branching, counting, and loop procedures in a Mathcad worksheet.

Mathcad X-Y Trace. A Mathcad utility that displays x and y values on a Mathcad graph.

Mathtype. A program from Design Science.com that is used to enter equations into a word-processing document.

Multiplication. A math process such as "$3 \times 4 = 12$" or "$a \times b = c$." Two types of multiplication are "sequence" and "polynomial." Two properties are "commutative" and "associative."

Multisim. A program from National Instruments Co. that aids in circuit and system simulation, using accurate device models and embedded test instruments and sophisticated graphing capabilities.

Non-real-time analysis. The signal is stored in memory and the analysis is performed at the speed of the computer, not at the same rate as the signal itself.

Normal distribution. The Gaussian probability density function of x from $x = $ minus infinity to $x = $ plus infinity. The cumulative distribution function CDF is the area under the curve from x_{min} to x_{max}.

Odd symmetry. The two sides $X(k)$ and $X(N - k)$ of a phasor spectrum have opposite phase.

One-sided sequence. A sequence in which all components are in the positive-frequency or positive-time domain. The sequence is constructed from the two-sided sequence.

Phase noise. Noise created by variations in phase of a signal. The rate of change of phase creates a phase noise power spectrum.

Phase shift network. An *RC* op-amp or DSP network that performs a negative $90°$ phase shift and a constant amplitude over a desired (e.g., speech) bandwidth.

Phasor. The complex exponential $A \exp(\pm j\omega t)$ is a phasor with amplitude A and zero average power. It can be at a positive or a negative frequency, depending on the sign of j. Two $\pm j$ phasors combine to produce a sine wave or a cosine wave at positive frequency.

Planck's constant. 6.63×10^{-34} joule-sec.

Postdetection filter. After RF/IF-to-baseband conversion, a signal can be filtered at baseband to improve the quality of the signal and can frequently improve signal-to-noise ratio.

Power spectrum. In an $X(k)$ two-sided phasor spectrum, the collection of phasor values (real or complex) at *(k)* from 0 to $N - 1$ is a *phasor spectrum*. The combination of phasors at $X(k)$ and $X(N - k)$ form a voltage or current *signal* at frequency (k). This signal has a power value, real (watts) or imaginary (vars), and a phase angle. The collection of the power values from 0 to $N/2 - 1$ is a positive-frequency (including dc) power spectrum.

Power (average). The average value of the product of voltage $v(t)$ and the current $i(t)$. If the two are in phase, the power is maximum and realvalued. If they are $90°$ out of phase, the average power is zero. The power value in a circuit can have a real component (watts) and an imaginary component (vars) and can have a phase angle θ with respect to some reference point.

Probability. A measure of the likelihood of an event. A tossed coin can be heads (50% probability) or tails (50% probability) for a large number of experiments.

Programming. Mathcad allows special program structures to be placed on a Mathcad worksheet. These programs greatly expedite and simplify certain kind of calculations that are difficult otherwise.

Pseudorandom. An event that is unpredictable in a short time interval but repeats at specific longer time intervals. Each occurrence may have random properties.

Random. An event that is unpredictable in time and frequency and amplitude.

Real-time analysis. An analysis that is performed in the same time frame as the experiment that is being observed.

Record averaging. A statistical averaging of many sets (records) of measurements of a noise-contaminated random signal.

Record length. The number of observations or measurements, from 0 to $N - 1$, in a sequence.

Sequence. A succession from 0 to $N - 1$ of values of a discrete signal in the time domain or frequency domain.

Sine wave, cosine wave. A pair of phasors, one at positive frequency and one at negative frequency, combine to make a sine or cosine wave.

Smoothing. The process of reducing the amplitude differences between adjacent samples of a discrete signal.

Spectral leakage. The variation of the amplitude of a discrete spectrum line at an integer value of $k \pm$ a small deviation $|\varepsilon|$.

Spectrum analyzer. An instrument used to view the spectrum of an RF signal on a CRT display.

State variable. The state of a system is its values of time, amplitude, frequency, phase, and derivatives at time (n) and frequency (k).

Statistical analysis. The properties of a noisy signal must be determined by procedures that extract an average result that approximates the properties of the noise-free signal.

Steady-state sequence. A sequence from 0 to $N - 1$ that repeats forever in the time $x(n)$ or frequency $X(k)$ domain. Each sequence consists of time, or frequency-varying components, possibly superimposed on a constant (dc) background. All transient behaviors due to initial conditions have decayed to zero long ago. Other methods for transient analysis are used (see the Appendix).

Symbolic. A method of problem solving in terms of variables that are defined not in numbers, but in math symbols.

System power transfer. In the frequency domain or time domain, the ratio of power out of a network to power into the network.

Time domain. Signals that are classified according to their occurrence in time t or $x(n)$.

Time scaling. A sequence of time values have a certain sequential relationship from the low end tothe high end. The maximum time minus the minimum time, divided by the number of time values, is the time scale factor.

Time sequence. An $x(n)$ time sample within a time sequence has two attributes, amplitude and position within the sequence, and $x(n)$ in this book is always a real number. A sequence has a positive-time first half and a negative-time second half.

Two-sided. A sequence from 0 to $N - 1$ is divided into the sequences 0 to $N/2 - 1$ and $N/2 + 1$ to $N - 1$. Point $N/2$ is usually treated separately.

Variance. The ac component of a complex signal. The rms value of the ac component is the positive square root of the variance.

Wave analysis. An algorithm to determine the properties of a signal. The properties include frequency spectrum, time waveform, amplitude, recordlength, period, power, statistics, harmonics, convolution, various transform values, and random properties.

Window function. A function such as rectangular Hanning, or Hamming that is used for windowing operations.

Windowing. A time or frequency record is multiplied by a window function that modifies the time and/or frequency properties of the record in order to make the record more desirable in some respect.

INDEX

Discrete-Signal Analysis and Design, By William E. Sabin
Copyright © 2008 John Wiley & Sons, Inc.

Technical Support

Contact PTC Technical Support if you encounter problems using the software. Contact information for PTC Technical Support is available on the PTC Customer Support Site:

http://www.ptc.com/support/

You must have a Service Contract Number (SCN) to receive technical support. If you do not have an SCN, contact PTC using the instructions found in the *PTC Customer Service Guide* under "Technical Support":

http://www.ptc.com/support/cs_guide

RECEIVED

APR 2 3 2008